NO REGRETS,
NO APOLOGIES

No Regrets,
No Apologies

The Life of Bob Rintoul

Bob Rintoul with David Finch

Kingsley

PUBLISHING

Cover and interior design by Dean Pickup
Project management by Kingsley Publishing Services
www.kingsleypublishing.ca

Front cover image: Bob with survey instruments, Medicine Hat, 1951

Printed in Canada by Friesens

2010 / 1

Library and Archives Canada Cataloguing in Publication
First Hardcover Edition

Rintoul, Bob, 1930–
        No regrets, no apologies: the life of Bob Rintoul / Bob Rintoul with David Finch. — 1st hardcover ed.

ISBN 978-0-9784526-6-7

        1. Rintoul, Bob, 1930–. 2. Businessmen—Alberta—Calgary—Biography. 3. Philanthropists—Alberta—Calgary—Biography. 4. Ace Explosives—Biography. 5. Calgary (Alta.)—Biography. I. Finch, David, 1956- II. Title.

HD9663.C32R56 2009   338.7'6622092   C2009-906395-6

First Paperback Edition

Rintoul, Bob, 1930–
        No regrets, no apologies : the life of Bob Rintoul / Bob Rintoul with David Finch. — 1st pbk. ed.

ISBN 978-0-9784526-7-4

Ordering information: www.kingsleypublishing.ca

All proceeds from this book go to the Bob and Nola Rintoul Chair in Bone and Joint Research in Southern Alberta, University of Calgary.

It gives me great joy to dedicate this book to my wife, Nola, who over the fifty-eight years of our marriage has been all things to me, but most important, my best friend.

Bob and Nola, fiftieth wedding anniversary, 2001. A three-day courtship led to more than fifty memorable and love-filled years.

Foreword

When I first laid eyes on Bob Rintoul he stood out in an audience of Calgary Flames season ticketholders. He sat in the front row. He asked tough questions and he expected sensible answers. He was pretty intimidating, actually. I did not know then that we would grow to have a wonderful friendship, an extraordinary philanthropic experience, and that he would, as he had done on so many other occasions with other organizations, become an important member of the volunteer Flames Ambassadors.

This book is a wonderful journey through his life. The son of a tough streetcar mechanic and union boss, Bob becomes an extremely successful explosives entrepreneur who sells his business back to his employees at the end of his career and starts another energetic chapter of philanthropy and retirement.

Born in Calgary in 1930, Bob walks us along the streets of a young Calgary—full of depression-era challenges and the promise of a Wild West city ready to embark on its own destiny to become a global business power. The adventure begins with escapades of youthful exuberance, a penchant for getting into trouble, and a passion for sports. You will also come to understand a loving mother and a strict Scottish father. Both shape his early years and start a lifelong search for peace with the man who taught him the hard way.

When the highly intelligent student opts for sports and girls in lieu of grades it does not take him long to appreciate the value and cost of choices. His lessons become yours and give you consolation and inspiration in your own life.

With Bob, you camp throughout Alberta and Saskatchewan with rough-and-tumble, hard-drinking seismic crews who were the forerunners of crews working rich oil and gas reserves. You bed down in shanty-town bunkhouses after rowdy nights in prairie bars. You survive dangerous winter roads and hole up in farm homes while blizzards blow through.

Bob introduces you to dozens of men and women who helped build a province. He weaves the genesis of their legacies through his chronological account of the dawn of a new western Canadian powerhouse.

Bob also tells a love story about his wife, Nola. Without sacrifice to propriety in the telling, he leaves room to see the joy of a young couple and their struggles, many relocations, and the fires that forge the strongest lifelong bonds.

After a career on the road in the pursuit of every opportunity, Bob recalls how the Rintouls cut their entrepreneurial teeth old school. Life savings on the line, he and other determined cohorts convince contemptuous Eastern bankers—who had yet to respect or appreciate their potential—that Western dreams were the best, and most secure, ventures around. Some investors were better students than others.

Chronicling the growth and success of his company in tandem with world travels and family events, Bob makes you feel more a part of the story than an outsider looking in. And when Bob and Nola turn their passions to philanthropy and volunteerism they do so with the same vigour they had already applied to raise their family, build their businesses, and to live rich and full lives. They teach us much about humility and generosity without conditions.

When you see this life in full, with "No Regrets and No Apologies," you will be left with no voids or ambiguities. As always, Bob puts it all on the table—a life laid bare for us all to see and learn from. You may not have the advantage of our friendship, but through his memoirs you too will get to know a great man and understand how to create legacies. Enjoy his story.

Ken King
2009

Contents

Acknowledgements

Over the years I have had the immense joy of making life-long friends and acquaintances. They have provided me with inspiration, happiness, and the opportunity to expand my knowledge and people skills. I would list the names, however I would be remiss if I inadvertently missed someone. You know who you are. I also dearly miss those who have passed away.

I have been blessed with a wonderful family: my wife, Nola, and my children, grandchildren, great-grandchildren, as well as my parents, grandparents, uncles, aunts, and first and second cousins. Over the years I have attempted to keep in touch with my extended family and they, as well as my immediate family, have given me the wherewithal to grow and to become who I am. Thank you.

I would also like to express my appreciation to historian David Finch and Charlene Dobmeier, of Kingsley Publishing Services, for their input and assistance and also to Nola for her many hours of editing.

I have always been capable of relating stories of my past. In the 1960s, many of my friends and acquaintances urged me to jot down these stories and eventually write my autobiography. Well, here it is, finally. If just one young person reads this missive and says, "If he can do it, so can I!" and grabs the brass ring, then my job is done. I have been blessed with a healthy, prosperous life, and have been allowed to live each day to its absolute maximum. The good Lord has been my guide and I have listened.

Part One

A City Kid

My name is Bob Rintoul. I was born in Calgary, Alberta, at the Grace Hospital at 5:30 P.M. on May 11, 1930, which happened to be Mother's Day that year. I weighed ten pounds, eight ounces and have been told by one and all that I nearly killed my mother. Be that as it may, Mother never had any more children. She was born Evelyn Vera Burke and married my father, Robert Parry Smyth Barrie Rintoul, on June 15, 1929.

That year saw the start of the dreadful Great Depression, which continued until the beginning of World War II in 1939. My childhood days were pleasant enough; I was unaware of the lack of money for "the extras" until I attended junior high school. My father was a Scotsman and very strict. Mother was a jovial, kind person whom I grew to love with all my heart. I loved my father, too, but in a different way. It is difficult for me to explain. My mother has always told me that he was very proud of me and loved me completely. But my father was never able to express those sentiments to me. He chose, instead, to scold me for minor irritants and spank me with an ironing cord for what he considered more serious misdemeanours.

When I left the hospital I was introduced to my first home. It was Hatfield Court, located on the Elbow River near the current Elbow River Casino, between Macleod Trail and First Street SE. We never lived in a house as such but rather in suites or apartments.

My first memory is from about age three. By then we lived in an upstairs three-room suite on Eighteenth Avenue between First Street and Centre Street East, near what is now St. Mary's High School. I could never figure out why my parents did not get up as early as me. Many years later my mother confirmed what I already

Robert (Bob) Rintoul was born in Calgary, Alberta, at the outset of the Great Depression.
This photo of young Robert was taken in 1932.

suspected. Dad was working six months on and six months off so that there was a job for all the employees at the Calgary Electric Street Car Railway. During his six months off there was no reason for my parents to rise early because we could afford only two meals a day. Dad was too proud to apply for welfare so he sold Christmas cards to provide us with some semblance of Christmas. He also repaired the roof of Knox United Church. If you have seen the pitch on that roof you would realize the courage it took to complete that task.

My dad worked for the Calgary Electric Street Car Railway, starting out as a painter and working up to mechanic on streetcars, buses, and trolley buses. He was very active in his union and eventually moved in 1955 to eastern Canada, where he became one of the founders of the Canadian Union of Public Employees.

We also lived in a suite on Fourteenth Avenue West between Eighth and Ninth Street. Here I remember playing with another kid from down the street. We would employ a card table and blankets to build a fort. This was also where my little dog died under the wheels of a passing car. My uncle Mid came in from the farm and loaded the dog in a gunnysack and took it to the farm for disposal. My dad advised one and all that I would never have another dog and he was true to his promise. I did not acquire a dog until after I was married and, up to my retirement, we owned several.

We also lived on Eleventh Avenue West between Second and Fourth Street. I have several memories from this location. One was of an old man down the street who made me a wooden trapeze man. It disappeared about thirty-five years ago but I found another and have it to this day. I have in the past and currently enjoy entertaining my grandchildren and great-grandchildren with this wonderful toy.

While visiting the Hays (close friends of my parents, and neighbours when we lived on Eighteenth Avenue), their son George, the older child, decided to put his sister Lois and me in a wagon and hook their Great Dane, "Pal," to the outfit. Unfortunately, Pal spotted a cat and took off like a shot. I vainly tried to apply the wagon brake but a power pole jumped up and pinned my arm between the wagon and the pole. George—fully aware of Mr. Hay's Scottish anger when aroused—chose to run away. They said I had a green stick fracture and I had to wear a cast for quite some time.

I also remember listening to the NHL hockey games on the radio with my dad. That was just before they reduced the league from eight to six teams. The longest overtime game ever played took place on March 24, 1936, between the Detroit Red Wings and the Montreal Maroons. Mud Bruneteau of the Wings, after 116 minutes and thirty seconds, in the sixth overtime period, scored the only goal, to win the game one to zero.

I also remember my mother going downtown to shop and leaving my sick father with a mustard plaster on his chest to relieve his congestion. However, he fell asleep and she was away longer than planned. When they removed the flannel and mustard plaster, Dad's chest skin, hair and all, stuck to the flannel. He was not a happy camper.

While living at this address I received my one and only tricycle. It was a beauty! Dad and his great friend Walter Quinn won it at the Stampede. It had a blue metal frame with whitewall tires. I was so proud of that trike. As you can imagine, toys were hard to come by in our home. One day a bigger kid stole the trike so my mom went looking for the culprit and the trike. When she found them, she gave the young thief a "skite oan the lug" (Scottish for a slap on the ear) and brought my trike home. You would not get away with that type of punishment today.

Life 101

In 1936 we moved to the best suite or apartment my parents ever rented, at 2121 Fifteenth Street SW, known as the Bowen Apartments, in the Bankview district. The apartment, one of four, was the lower north unit. It consisted of an east-facing, enclosed glass sun-porch, living room, Mom and Dad's bedroom, bathroom, dining room, kitchen, a bedroom of my own (the very first and last while I lived at home), and a back porch. The rent was twenty-five dollars per month and the landlord told my dad he would never raise it. Dad took this as Gospel.

I have a number of fond memories of living in this great apartment. The people who lived in the other downstairs apartment were the Grouses. They had two boys, Dale and Dalmer, who were great playmates. Their dad worked for the Ford Motor Company, which employed rubber model Ford cars as an advertising gimmick. The boys owned a number of these and gave me a 1934 or '35 sedan, painted tan.

It was at that time that my dad started playing golf, and when my parents gave me a Kodak box camera for my sixth birthday my first photo was a picture of Dad with his golf clubs. The reason I remember this occasion was because of the scowl on Dad's face. He felt I was not taking the picture "just right." This was the first of a continuing pattern of incidents with my dad that continued into my teens.

When I was young my hair grew in all directions and at about five years old Mother made me a "stocking cap" to put on my head each night while I slept. The cap was made from an old silk stocking with the foot cut off with one open end knotted and, voila, you had a "stocking cap." It worked very well and after a couple of years my unruly hair was tamed.

My first school was Bankview and I have happy memories of those times. The King and Queen made a cross-Canada trip in 1937 when I was in grade two. The students walked from the school on Seventeenth Street and Seventeenth Avenue to Fourteenth Avenue and Fourteenth Street, where we lined up to observe the parade.

I experienced my first heartthrob while we lived in Bankview. Her name was Eleanor and she sat in front of me in class. We eventually were reacquainted in about 1960 when she accompanied her husband to the Doodlebug Golf Tournament. She subsequently developed serious medical problems and dropped out of our circle of acquaintances. Sadly, she passed away in the 1990s.

Not only did I start my scholastic education at Bankview School but also my basic education in "Life 101." One of the students purloined some condoms from his parents' dresser and quite a few fellow students had great fun filling each of them with about ten gallons of water. I did not know what a condom was used for until I moved to the Mission District. Those kids in Mission knew everything!

This was where Dad taught me to be more aggressive when confronted by adversaries. The Protestants and the Catholics seemed to dislike each other, something about going to separate schools and probably what our parents taught us. The Catholics would separate me from my friends and pick on me. One day Dad was walking home from the streetcar stop after work and saw this performance. These kids were somewhat older and he cautioned me: "If somebody is bigger and older than you, get a stick or whatever, whittle them down to your size, and beat the crap out of them." I never forgot that lesson.

It seemed that most of the Catholic kids in our neighbourhood were older and bigger than the boys I chummed with. With this problem constantly confronting us, we formulated a plan. One of our gang's dads worked for the Orange Crush Company. Orange Crush was a popular soft drink in those days. It was in a brown bottle with circular ridges and had the name emblazoned on the bottle with an orange logo. These bottles were for advertising purposes and Bill's dad had a supply of a sticky orange dye that they used to fill bottles, which simulated the real Orange Crush. We had stored a few bottles of this stuff on the roof of their garage. One day, as our antagonists chased us down the alley we turned the corner and scrambled to the roof of the garage. When our larger opponents came to pull us down and beat the tar out of us, we poured the dye on them. Their hair turned orange for about two weeks. That was the last of the feud; they never bothered us again.

When I was a kid I never questioned this conflict between the Catholic and Protestant kids but after some investigation I realized that the Protestants and the Catholics had been fighting for centuries in Great Britain. As my dad was born in Scotland I assumed he arrived in Canada with this attitude. I am certain that if I had married a Catholic girl she would not have been welcome in my parents' home.

During my junior high school years I was fortunate enough to be selected for an "A" hockey team, playing shoulder-to-shoulder with others of the same calibre from diverse backgrounds. Some of these teammates were Catholic, and we became good friends. There's nothing like team sports to solidify a friendship, and to this day I have a host of great Catholic friends and acquaintances. I believe that the separate school system encourages these kinds of religious differences.

There was a vacant lot directly north of our apartment and due to the low contour of the landscape it collected a lot of runoff from the snowmelt in the spring. It provided a great body of water for rafts and accommodated self-made toy boats. On the way to school the ploughs piled the snow in the gutters and the melt would run underneath. It was great fun to walk on the drift and try not to fall through. Of course, you nearly always did and would have an overshoe full of water. Mothers were not impressed.

We played road hockey on Twentieth Avenue SW between

Fourteenth and Fifteenth Street in front of Dr. Fish's house. We used road apples for pucks and rocks for goalposts. Road apples, for the uninitiated, are frozen horse poop. The road surface was oiled and because of the uneven nature of the surface the depressions caught the snow that hardened and made pretty good "ice." Some of the guys who played with us were Fred Hilderman, Teddy Walshaw, Billy Aeirdale, and Larry Fish.

Down the alley from us, on the way to school, was Jimmy's house. One winter his dad built a rink that was lit at night and had a goal placed under their kitchen window. I knew a good thing when I saw it—this was better than playing with road apples on the road. I continued to play one evening after Jim exited for dinner. My dad had made me a wooden puck, as rubber pucks were too expensive. I reared back, fired the puck at the net, and drove it through their kitchen window onto the table. Jim's dad was English but I guess he had as bad a temper as a Scottish dad because he came roaring out of the back door. I ran like hell down the alley and as I turned the corner by the caragana hedge there was my dad (he had somehow caught wind of this foot race between a promising young hockey player and an irate broken-window owner). He yelled, "Get in the house!" I peered out the kitchen window and by now Jim's dad had rounded the corner by the hedge, where he ran into my father, who grabbed him by the front of his coat and said, "If I ever catch you chasing my son again I will beat the hell out of you. I'll pay for your goddamned window." When Dad returned, all he had to say was "Do not play in Jim's yard again."

There was an old sandstone quarry at the corner of Twenty-first Avenue and Fifteenth Street that provided an ideal playground. It was actually a large excavation about twenty-five hundred square feet and thirty feet deep. We would place boulders on the floor of the quarry and throw large rocks from the top and smash the boulders. One evening as Fred and I were playing this game he was at the bottom, repositioning a rock. I was above him holding a huge rock in my arms extended over the edge of the cliff. I shouted down, "Fred, I can't hold this any longer!" and let it go. Fred was a little slow at the best of times and the rock caught him squarely on the head, knocking him unconscious. I looked down, panicked, and ran home. I never said a word about it. A couple of days later Fred showed up at school, apparently none the worse for wear. What did I know? I was only about eight years old and scared as hell. My parents never found out about it for nearly ten years. Fred and I ended up in high school together and he went on to be a pretty good doctor. Maybe the blow helped ... To this day, I have never figured out why I had not just stepped back with the rock and allowed it to fall harmlessly to the ground. But eight year olds do not always think straight.

One very cold winter night my dad sent me on an errand to the "wee" store on Seventeenth Avenue and Seventeenth Street for a can of tobacco. It was nearly dark and he had given me a quarter to buy the item. For safekeeping I placed the quarter in my mitt but could not find it when the time came to pay for the purchase. I was terrified of what Dad would do to me for losing the quarter and told the storeowner and a customer of my dilemma.

The storeowner offered to give me the can of tobacco and I could return with the money. I hesitated, as I knew the consequences when I returned home and told Dad. The customer, obviously a kindly gentleman, paid the quarter for me and advised me to say nothing. Of course all this had taken considerably more time than a normal trip to the store and my dad was curious about what had taken me so long. I tried to bluff it through but I never was a very good liar. So out came the ironing cord for a whipping across my backside as a lesson: "Do not accept money from strangers."

One summer, when I was about six or seven, my mother and Grandma Burke took me to the Calgary Stampede. I wandered off and after a lengthy hunt they hopped on a streetcar and returned home to make Dad's supper. My mother was torn between finding me and having my father's dinner ready for him after his hard day's work. You did not have to worry about pedophiles as you do today, and Mother knew I could find my way home. The Stampede grounds were in the same location as they are today on Seventeenth Avenue and Second Street East, and we lived at Fifteenth Street and Twenty-first Avenue West. That was a long ways for a little kid to walk. So I walked to our friends, the Hays, on Eighteenth Avenue and First Street East, borrowed a streetcar ticket, and rode home. I do not remember the discussion between my dad and mom but I have been told the word "divorce" was used. My dad had wanted a son so bad when I was born that my mother has often said, "I do not know what Dad would have done if you had turned out to be a girl, probably divorced me." I was his pride and joy, but again, he did not—or could not—tell me so.

One winter day Fred Hilderman and I were sledding on a hill near Fourteenth Street and Seventeenth Avenue. We stayed out for hours in the bitter cold. When I returned I had no feeling in my toes; they were frozen. Dad immediately went outside and brought in a basin of snow and rubbed my feet for a couple of hours. Man, did they hurt when that freezing came out! This action proved that my dad loved me but this did not register in my young mind. (As it turns out, rubbing with snow is a no-no. Use heat!)

One summer day Jimmy Condon's candy store on the corner of Seventeenth Avenue and Fourteenth Street caught fire. Fred and I heard the fire engines and ran to the fire with Fred's dog loping along behind us. The candy in the windows melted into big blobs, the roof caved in, and there was not much left an hour later. On the way home Fred's dog was run over by a car and we had to carry home a dead dog. Life was cruel but we learned many lessons from those early experiences.

Once, intensely interested in watching workmen remove a wooden form near our apartment as they repaired a portion of sidewalk, I moved in a bit too close to the action. When they lifted and swung the form, the pointy end of a big nail caught me squarely between the eyes. I ran home, blood streaming down my face and clothes. My mother very calmly cleaned me up and bandaged the wound. I still remember how painful that wound was: like eating an ice cream cone on a very warm day as fast as you can.

About a block away from our place was a kindly elderly woman who made watermelon pickles each summer. She only used the rind of the watermelon and I was at her place to clean up the watermelon meat. I ate the entire watermelon. Man, was I sick. Luckily, it never affected my love for the fruit. To this day I look forward to watermelon season.

My aunt Kay was in training for a nurse while we were at the Bowen apartments. She had recently undergone a mastoid operation but wished to attend her graduation. Recuperation was very painful and we attended to give her support. I remember the little sandwiches and sweets. At the time that Aunt Kay graduated, jobs for registered nurses in hospitals were few and far between. Remember, this was the depression. She stayed with us and we shared my bed. She eventually got a job as a specialist for a live-at-home patient, which required odd working hours. I remember one Easter, Aunt Kay hid a huge Easter egg, all decorated on the top with candy icing and full of little chocolate eggs. That was the neatest Easter I can remember.

Auntie Jean (Dad's sister) married Uncle Jack Ferschweiller (her second marriage) and the reception was held at our home—pretty neat for a kid of eight or nine. A large crowd of people attended and there were various delicious sandwiches and desserts available.

My tonsils and adenoids had been removed recently and my mother cautioned me not to vomit while in the hospital. Throwing up there was not an option. You did not want to soil the hospital's blankets! In those days appearance was everything. I was careful to heed her advice, to my (and my grandma's) detriment. After the operation Grandma Rintoul insisted I stay with her. She lived in an apartment block on the south side of Sixteenth Avenue between First Street East and Centre Street. Of course, Grandma spoiled me rotten, giving me all kinds of ice cream and candy, and I threw up all over her new blankets.

I have pleasant memories of ice cream. I was addicted to "thumb sucking." My parents attempted all forms of cures to no avail. These included covering the thumb with bandages, which I would rip off in my sleep, and hot pepper, which I would eat, etc. Finally, they bribed me with a nickel ice-cream cone each week. Problem solved, and I've liked ice cream ever since.

The Bankview location was also the breeding ground for my early childhood medical problems, which included mumps, chicken pox, and bronchitis. The Burke side of the family has congenital bronchial problems as far back as my grandma Burke. Of course, I picked up all three problems as a kid.

Union Man

The winter of 1938 was one of the coldest winters ever. The temperature dropped to minus 48 degrees Fahrenheit with blowing snow for about ten continuous days. My dad at that time worked as a mechanic for the Calgary Electric Street Car Railway. The cold weather caused havoc for streetcar traction. Streetcars were off the tracks at numerous locations. Dad and his fellow workers were required to grovel around in the snow using large steel plates and the wrecking car to slide the cars back on the tracks. During one three-day stretch Dad never saw his bed and caught catnaps when possible. He became very ill as a result of this ordeal and was bedridden for days. My father worked hard for his money and during the depression was lucky to make thirty-five cents an hour.

The Street Railway Sick Fund (there was no insurance or payment for sick employees at that time; each employee funded the plan with monthly contributions) held a Christmas party each year for the children of the Calgary Electric Street Car Railway employees. It was a major event in my life. They gave out a bag of candy, a mandarin orange, and a gift. Santa Claus was the fattest employee, dressed in a Santa suit.

Dad started to work for the Calgary Electric Street Car Railway in 1928. From the paint division, repainting cars with lead paint, he progressed to a mechanic and then the depression hit. The union worked out an arrangement with the City of Calgary so that the employees worked six months on and six months off, allowing everybody to earn some yearly income, but at a very minimal wage. This arrangement continued until about 1936. Because Dad was not working full time, the union did not reinstate his seniority from 1928. He fought them and was able to get his seniority reinstated. He was a crusty Scot—small but tough.

His first position with the union was as a steward, and in 1947 he was the first shop employee to be elected president of the Amalgamated Association of Street and Electric Railway Employees, Local #583.

From 1947 to 1956 he was elected or appointed to the following: president of the Calgary Federation of Public Employees, Calgary Trades and Labour Council, and the Alberta Federation of Labour. He also served as vice-president of the Calgary Community Chest (currently the United Way), and was on the boards of the Calgary General Hospital, Calgary Library, Alberta Red Cross, the Alberta General Education Curriculum Committee, and was a member of the senate of the University of Alberta.

Sometime during the early 1950s, Don McKay, the mayor of Calgary at that time, offered Dad a job with the city as labour relations manager. The salary was considerably more than he was making. I remember him coming home and discussing it with

Mom. Finally, he said, "No, my union people need me."

All his union endeavours were aimed at improving the welfare of the union members. But when push came to shove, his union buddies let him down. Sometime in the mid 1950s he attended a Canadian Congress of Labour convention in Regina. I was party managing a crew at Assiniboia, Saskatchewan, and drove up to spend a day with my parents. He was up for election to one of their board positions, likely vice-president. He was assured of the support of his western Canadian union buddies, but in the end they let him down for their own gain.

It was such a disappointment, not so much the defeat but the treatment he received from his so-called buddies. He wept. I never forgot seeing him cry (the first time in my life). That situation taught me a valuable lesson: keep your back covered at all times. However, my dad's integrity carried him through. He continued to help the union members achieve their goals.

In 1956 he moved to Ottawa to become Canadian director and a founder of the National Union of Public Service Employees. In 1963 he was instrumental in joining with the National Union of Public Employees to form the Canadian Union of Public Employees (CUPE). He was their first national secretary-treasurer until he resigned in 1967.

While in Ottawa he also served as co-chairman of the Religion-Labour Council of Canada, a member of the National Committee on Unemployment (Canadian Labour Council), and vice-president of Court of Referees-Unemployment Insurance (Ottawa and Calgary). After his retirement in 1967 due to stress from his work at CUPE he became business manager of the Ottawa Hospital Union, Local #576, and Canadian delegate to the World Congress of Labour.

I grew up in a union home and in many ways it defined who I became in later years. My relationship with Dad was never a bed of roses. We seldom saw eye to eye on any subject but I eventually recognized how intelligent, proud, and committed he was to a fault. When I was about ten or eleven, he said, "Bob, if you're smart, you'll move into management, and forget blue-collar work." Then he said, "You'll have negotiations with unions, but management strength is stronger than union strength." I thought that was pretty good advice, especially coming from him, and attempted to position myself to attain management positions.

The union members would sometimes have poker parties at our home and I was privy to their conversations. As a result, I was able to understand what turned the working man's crank, what he wanted from a financial point of view, what his dreams were. The working man wants to be treated with integrity and respect. That was a great education for me and allowed me to practise these lessons when I became a party manager on a geophysical crew and a company owner. I attempted to treat my employees according to what I had learned, though sometimes this was difficult because the company umbrella of wages and benefits limited my options.

When I was able to own a company, the employees received top remuneration, including benefits. I even paid their CPP, but that became a tax problem so we dropped that benefit. Their health and dental, including orthodontics, were covered as well and they received 15 percent of the company profit before taxes through a federal government approved profit-sharing plan. They knew how well the company was faring financially and it became a very successful operation.

Family Ties and Country Visits

My grandma Rintoul was a wonderful grandma. I was her first-born grandson and she spoiled me rotten. She lived at numerous locations in Calgary including a suite on the south side of Fifth Avenue West between Fourth and Fifth Street. It was here that I first had Grandma's favourite meal: frenched pork tenderloin with mashed potatoes and spinach and a fried egg. This was followed by chocolate cream pie, which was my dad's favourite.

On the north side of Eighteenth Avenue East between Second and First Street, I remember playing jacks with my older cousin Betty on the top landing of the staircase. My grandma lived on the upper floor of a rooming house on the south side of Twelfth Avenue East between Third and Fourth Street (this is currently the main entrance for the Stampede Casino). Grandma was a prolific knitter and when I was about ten she knit me a Toronto Maple Leafs sweater complete with a big maple leaf in the centre of the chest. I was extremely proud of that sweater.

Grandma also worked as a chambermaid at the St. Louis Hotel. As a little fellow I would accompany my mom on a visit to Grandma at this location.

Grandma eventually moved to the Cariboo area of BC to look after some rancher's children. Then in the early 1950s she moved to Vancouver on Twelfth Street just off Granville. Whenever I visited Vancouver on business, she would invite me for supper. She would offer me a drink of scotch to whet my appetite. She requested my choice of mix and I asked for "a little water." She filled a tumbler almost to the top and splashed in a small amount of water. I said, "GRANDMA!" She replied, "What? Have I put too much water in it?"

She eventually moved to a home for the elderly on Harrison Drive where I continued to visit her. Grandma was either knitting or buying something for me. On my fifteenth birthday, she gave me one of my most cherished possessions, an engraved identification bracelet. I still have it. She died at age ninety on October 13, 1969. Her full name was Elizabeth Nelson Rintoul. I loved her very much.

As a child and young man one of my favourite pastimes was visiting Grandma and Grandpa Burke at their farm on Symon's Valley Road. Travelling to the farm was a major undertaking for our family sans a vehicle. From whatever part of Calgary we were currently residing in we took streetcar transportation on the

Tuxedo route to the "loop" at about Twenty-sixth Avenue and Centre Street North. At this point we either waited for our ride to come from the farm or started walking until they met us. The route from the loop was north on Centre Street past the water spout (there was a well with a twelve- or fifteen-foot standpipe and a piece of fire hose attached so people who did not have water in their homes could get water) to Thirty-second Avenue NW, and west on the north side of the creek (no longer there) to Fourth Street NW and out to the farm. It was located where the present Sir John A. Macdonald School sits (just north of Fourth Street and Sixty-fourth Avenue NW), which I think is rather fitting considering the history of the Burke family.

The farm was a great place for a city kid. Usually we celebrated our Christmas at the farm, joining a number of our relatives, including uncles, aunts, and grandchildren. The two uncles who lived on the farm (Gordon and Mid) left shortly before dinner to milk the cows. The table was laden with turkey and all the trimmings. After dinner the living room would be opened up (it was kept closed most of the winter because the house was heated by coal and it was poorly insulated). A singsong followed, with Aunt Nellie playing the piano accompanied by Uncle Mid and Grandpa on violins and Uncle Gordon on guitar and mouth organ (he had a mouth organ holder that fit around his neck and he could mouth it while playing the guitar). After the singsong a dance was in order. The evening festivities carried on until the wee hours of the morning.

If we stayed overnight, which was not often since generally Dad had to be at work early in the morning, we would retire to one of about six unheated upstairs bedrooms. When you hit the deck in the morning it did not take long to get into your clothes and head downstairs to a big breakfast of porridge, bacon and eggs, and toast and coffee.

Visits year-round were the norm, including Sunday fried

chicken feeds. Uncle Mid would catch a chicken, which, having its head cut off, would run around in circles for what seemed like forever with its lifeblood spewing out of its neck. As a child this phenomenon fascinated me, and at that time nobody gave me a satisfactory explanation. There were always lots of old cars, trucks, and farm machinery to play on and animals to watch as the men fed them. It amazed me that my grandpa, who was blind, could feed the pigs and milk the cows.

From about age eight to twelve, I spent my summer holidays at the farm. I asked Grandpa how he knew it was his property. Grandma made us a picnic lunch and we walked out into the field where he found a metal Canadian Government Survey post. He explained how they engraved Roman numerals on the rod to identify the land in mile-by-mile squares, with a road allowance every mile east and west and every two miles north and south. The land was surveyed into sections, townships, ranges, and meridians. That stuck with me and when I became a surveyor I often reflected on the day when Grandpa Burke taught me how western Canada was defined and surveyed.

My grandfather was a wise old man and he counselled me as a youngster to avoid the lure of greed. He told the story of how he lost two fortunes in the real estate business due to greed.

Bob, age seven, at his grandparents' farm, with Grandpa Burke's hat and cane, 1937. Notice the patch between the eyes where Robert suffered a puncture from a nail when he got too close to a construction crew.

30

Mom would often go out to help with the harvest. It was a time of frantic work to harvest the crops before the frosts, and any and all help was gratefully accepted. The crops were previously cut and bundled with binder machines, then the bundled grain was stooked. This was a process of putting one bundle or sheaf in the middle and four or five around the perimeter of the first. This allowed the grain to shed water and also assisted in drying the grain.

The threshing crew arrived complete with a threshing machine, where bundles of grain were forked into a hopper and an interior separator separated the grain from the straw. The grain flowed into a wagon and the straw was blown into a huge pile. The bundles were transported to the thresher by a wagon pulled by two horses that travelled up and down the rows of stooks where men pitch-forked the bundles onto the wagon. The wagon driver placed the bundles in an orderly manner until the wagon was piled high and then driven to the thresher. One of my big thrills was to accompany the bundle wagon and be allowed to drive the team as the wagon was loaded.

The men sat down to a huge lunch set up adjacent to their steel-tired wagon bunkhouse. Long tables and benches were laden with food. The women—Grandma, Aunt Nellie, my mom, and any other available help—spent days before the harvest gang arrived preparing pies, cakes, homemade bread, etc., to feed the famished crew. There was an abundance of meat, potatoes, garden vegetables, milk, coffee, and pies.

If the threshing took longer than one day there was an evening meal served the same way and another steel-tired wagon complete with sleeping bunks for the crew. Usually, there was some music at night performed by threshers, who often brought along guitars, violins, and mouth organs.

Then next morning it was breakfast with oatmeal porridge, bacon, eggs, toast, jam, and gallons of coffee in big blue enamel coffee pots. The food was cooked on a wood and coal cook-stove with a water reservoir attached to heat water. The men washed up at a long plank set up outdoors with enamel washbasins, soap, and towels. Those were good days for a little boy. To this day I still relish the sounds and smells of harvest time.

One time while at the farm with my mother I decided to follow my uncle Mid while he was cultivating a field with a two-horse plough. The family dog and I followed that plough all day. That night I became deathly ill with a high temperature and was in a semi-comatose condition for well over a week (in those days going to or having a doctor come out was out of the question—no money or phone). I finally came out of it and to this day I have very weak ankles that I did not have prior to the bout of "sun-stroke." Some have said I may have had a brush with polio. We will never know, but I was seriously ill.

One of the small pleasures I had at the farm as a little guy was going to the chop-house (a single wall barn about sixteen feet by sixteen feet where they chopped the grain for pig feed). I would close the door with only the sun shining through cracks to partially light the area. There I sat quietly with a hammer and when the mice started to scurry about me I would smash them. Not everyone's idea of a good time, perhaps.

It was always a great adventure. On one occasion when I was about four I followed a kitten into the chicken coop. These kittens were wild and wanted nothing to do with a human. When it crawled into a little hole I followed, but I became stuck in the opening. I reached my hand farther into the hole to grasp the kitten and it bit through my index fingernail. Fortunately, my screams of anguish brought fast action from the house.

When I was a little older I was allowed to go to the farm for my summer holidays. Sometimes my cousin Howard Burke from Edmonton would join me. The farm was no place for lazy boys; you were expected to pitch in and help. When I was smaller I collected eggs for Grandma and emptied the slop pail of potato peelings, refuse, and used wash water. The chickens made short work of these goodies.

Not only were you expected to earn your keep but any waste of food was a sin. When I was about six, while eating breakfast, I over-salted my boiled eggs. Knowing that I would be scolded for leaving my eggs, I added sugar, thinking that would solve the problem. Of course, that made it worse but I knew I had better finish eating those eggs.

We would rise at 5 A.M., round up the milk cows in the pasture, and drive them into the milking barn. There were about forty-five animals. We would close their stall yokes and drop their feed from the loft. Grandpa and both uncles would milk the cows by hand into gallon pails then carry the milk to the attached milk shed and pour it into gallon cans placed in cold water.

When we were finished Howard and I were required to "strip" cows, that is, squeeze the last bit of milk from the teats. If either uncle squeezed additional milk from a stripped cow it was head-first for us into the water-filled horse trough.

The haying season gave Howard and me another opportunity to work with our uncles. A two-horse sickle mower was employed to cut the hay and then it was left to dry for a day or two. A one-horse dump rake gathered the hay into piles. (They called it a dump rake because the machine had half-round steel tines behind the unit and when the tines were full, the operator would kick a lever and the machine would "dump" a pile of hay.) A two-horse hitch hay wagon lined up beside a stack and we would pitchfork the hay into the wagon. Our job was to stomp the hay down so more could be piled on top.

When the wagon was full someone drove it to the barn where a door high up the wall allowed access to the loft. One person would pitch hay into the loft and two or three others in the loft would pile it as neatly as possible. Howard and I, as the youngest, were given the back wall position where there was very little air circulation. It was dusty and hot and the thistle-infested hay laid scratches on our shirtless upper bodies.

I also remember watching my uncles break a horse. They would harness the horse to a full grain wagon and drive it up and down the alley where the cattle came in from the field. Then they unhooked the horse from the wagon and put on the saddle. My uncle would carefully step into the stirrup and in a very soft voice attempt to calm the animal. We then had a rodeo, which was very entertaining for us wee boys.

Every summer Grandpa Burke (blind and all) would take Howard and me to Bowness Park, owned and operated by the Calgary Municipal Electric Railway. Someone would drive us from the farm to the Tuxedo Loop where we boarded a streetcar to the centre of Calgary and then transferred to a Bowness Route streetcar. The ride was of such length that the car stopped at the Shouldice Bridge, which crossed the Bow River, and each passenger had to pay an additional fare. On occasions when the Bowness Route was very busy a trailer car was attached.

Bowness was an oasis during the depression. Its popularity continued into the early 1960s at which time more vehicles and entertainment options became available to residents. However, with the current increase in immigration it allows citizens with fewer resources a reasonable day of entertainment. You can currently travel to Bowness using a city transit bus.

The park offered varied entertainment including a series of lagoons complete with rental canoes. There was a lake with an island that housed a building that supplied power for a large fountain. The fountain spewed a thirty-foot column of water, lit at night by varied coloured bulbs. The building also contained the facilities for music, which was projected by large speakers.

There was a carousel, horseshoe pitches, and a large concrete pad complete with a painted checkerboard where you moved your checkers with a metal hook. Concessions dispensed hot dogs, hamburgers, fries, ice cream, and candy, including all day suckers (five cents). Grandpa would give each of us a one-dollar bill—a lot of money in those days—and tell us to return at a specific time. Both Howard and I looked forward to that day each summer and we came back happy and full of junk food. If we were five minutes late, Grandpa would scold us. We could never figure out how a blind man without a Braille watch could tell when we were five minutes late. To this day we are no wiser and when we questioned him he simply said, "That's my secret."

While staying at the farm we also helped Grandma on Friday nights gather and candle sixty dozen or so eggs for sale on Saturday in Calgary. Most of current society is unaware of this process, so I will explain. You hold an egg between your thumb and forefinger up to a lit candle. Light from the flame actually shines through the eggs and you can see inside the yolk. If you see a red or any spot in the yolk, you discard the egg. Spots in the egg mean it has been fertilized. Back in the old days, when the hens and the roosters travelled together, you sometimes had a fertilized egg. Nowadays the hens are kept separate, and the rooster is not allowed near the hens.

On Saturday we would accompany Grandma and one of the uncles to Calgary. While the uncles were picking up machinery parts, Grandma and Howard and I would deliver the eggs and chickens to steady customers living in little brick row houses on Centre Street between Fourth and Sixth Avenue. We then went to the market, where Olympic Square is currently located, and bought the necessities for the following week that the farm could not provide. In 2001, in remembrance of these times, Nola and I and many others contributed to a larger-than-life size bronze statue located near the Ranche Restaurant in Calgary's Fish Creek Park. The statue is named "Egg Money" and the contributors have inscribed an individual pioneer woman's name on one of many bronze plaques. I dedicated my choice to my grandmother, Edith Louise Burke (Jordan).

In the summer of 1941, my mother and I caught a ride with a

salesman to Lethbridge to visit Aunt Grace, Uncle Howard, and my cousin Blair. The highway was gravel, and I was excited to be travelling out of Calgary. I had never been to another city in Alberta. It was near the end of the depression and people were still struggling to make a living.

Aunt Grace and Uncle Howard owned a farm in Taber, but due to the dry conditions they could not make a go of it on the land. Uncle Howard had taken a job in Lethbridge as a mechanic at a downtown service station. One day I visited him at the garage and he took me to Kresge's department store restaurant for my first root beer. It was a hot day and I can still remember that cool sudsy root beer. They lived in a second level suite in a house and was it ever hot! No air conditioning in those days.

We also occasionally travelled to Trochu to see Aunt Jean, Uncle Jack, and my two cousins Betty and Lucille. One Christmas we started out on the bus but it broke an axle near Acme and we were held up for some hours until a replacement bus arrived from Calgary. Both of my cousins were telephone operators. Up until the early 1950s, most small towns and all country areas were not equipped with direct dial telephones. Rather, each line had a number of telephones on one circuit, called a "party line." When you wanted to make a call, you rang the operator and she rang the number you wished to contact. Of course, all those on your party line could listen in on your conversation. (There were few secrets in those days.) My operator cousins kept their parents advised of our whereabouts and estimated arrival time.

On another trip to Trochu in the summer of 1940, I learned to ride a two-wheel bike. It was one of my cousin's bikes and I would rip down the main street, which had a fair decline, at great speeds. After a few complaints from the locals I was banned from the main street. I met the Purvis brothers at Trochu: Al became a member of the 1952 Canadian Gold Medal Olympic Hockey club—the Edmonton Mercurys, who were the last Canadian team until 2002 to win gold. The Edmonton Waterloo Mercury dealer sponsored the team.

The "hangout" for the young folk in Trochu was the soda bar in the local drug store. The soda bar was about thirty feet long and varied ice cream and soda delights were available. My favourite was and still is a "black and white." It consists of a two-scoop sundae topped with chocolate sauce on one scoop of ice cream and marshmallow sauce on the other. My uncle Jack was the BA (subsequently Gulf) gas and oil agent as well as the John Deere agent. He was also the mayor of Trochu. It is interesting to note that he hauled the fuel from Turner Valley in his own trucks.

City Life

But back to city life. While we were living in the Bankview area we would board a streetcar for Bowness Park. It was a big outing for me and as previously explained there was a lot to do at the park. Mom would pack a picnic lunch and off we would go, early in the morning, returning after dark. I usually slept on the way home. On one of these excursions we placed a big water-melon in the river to cool, packing stones around it (the Bow River was the northern perimeter of the park). The Ghost dam farther up river released water every day about 4 P.M. and our water-melon started down river! Fortunately, Dad rescued it.

Some of our early entertainment was pretty simple. We played tiddlywinks, a game where each player used a large plastic disc, which was the "shooter." You placed it between your thumb and forefinger and attempted to put a smaller same-coloured disc in a small cup by forcing the big disc onto the small disc that rested on the floor, more or less flipping it into the cup. Each player had five small different-coloured discs and the first one to get all five in the cup, won. Other games were pickup sticks and card games.

Medical and dental care was quite rudimentary in those days. There was no money for necessities other than food, a roof over your head, clothing, and basic leisure activities. Our dental care was taken care of through the school. When we had a cavity we were sent to a dentist hired by the City of Calgary, with an office in the tower above the old sandstone City Hall, still in existence. This dentist did not use freezing of any type. He drilled away with the equipment of the day. The drill rotated very slowly, and the pain was excruciating. After a couple of these torture sessions you made the decision to let your teeth rot until they needed extraction. Still no freezing, but less painful.

In 1940 the landlord at the Bowen Apartment raised our rent from $25.00 to $27.50 per month. My father was beside himself because the landlord had promised he would never raise the rent. Never is a long time and my dad moved us to new quarters in the Mission district.

Because I was big for my age the older kids in Bankview used to pick on me. I was not really much of a fighter nor did I wish to engage in fisticuffs. But the kids in the Mission, Parkview, and Parkhill areas were of a different breed. They would as soon beat the crap out of you as look at you. When we moved there, I had to become a lot tougher. I continued going to Sunday school at Scarborough United Church, which was adjacent to the Bankview district. One Sunday after church I ran into one of the older kids from Bankview who used to beat on me. I calmly took him apart. The word went out: "Rintoul's not a patsy anymore."

The address of the house we moved to in the Mission District was 2416 First Street SW. We had an upstairs suite consisting of three rooms: a kitchen, a combined living and dining room, and Mom and Dad's bedroom. A single gal who worked at the Independent Biscuit Company occupied the other two rooms on the upper floor, and the Laughrens—Mom, Dad, and sons, Graham and Grant—lived on the downstairs floor. We all shared a bathroom that contained a sink, toilet, and old-fashioned tub.

I slept on the chesterfield in the living room. After our lovely apartment on Fifteenth Street this was a real "come down." However, as a ten year old I do not remember being resentful or unhappy over this change in living accommodation. I imagine, as are kids today, we were pretty resilient and took things as they came. Our suite was also the access to the attic and we used that for storage. The attic floor consisted of "stringers" with lath and plaster applied for the ceiling below. We had 1" x 8" boards to walk on, but it was pretty unstable. One time Dad decided to re-paper the living room ceiling and after completion went up to the attic to put away the leftover materials and tools. He fell through the ceiling. I was in the living room and started to laugh. My father was not impressed with my mirth and yelled at me to "get him out of this %&#@*^% place," which I did after he promised he would not scold or spank me. I pulled him out and he was true to his word, but he was sure angry. He then had to cement the ceiling because the laths were broken and then re-paper. You could see that cemented spot through the paper thereafter.

My dad was usually true to his word, particularly when it involved punishment for one of my real or imagined misdemeanours. During these years of my life, six to about twelve, I really looked forward to the Calgary Stampede's Wednesday "kids" day. Children from Calgary and surrounding districts were entertained with a free grandstand show complete with prizes including a horse and saddle, puppies, bicycles, etc., and all midway rides at a reduced price. This was my day and I could hardly wait for the occasion each summer. Only birthdays and Christmas topped the Stampede.

In the month of May, I apparently lied to my dad and he felt a fitting punishment was "not to allow Robert to go to the Stampede." I figured this was such dastardly treatment that he would relent and allow me to go. I was on pins and needles for about two months. The morning of the big event I got up early as usual on this auspicious day and got dressed to go the "THE STAMPEDE."

Bob, 1937, and his parents, Evelyn and Robert. Bob's father, a Scotsman, was a stern disciplinarian.

However, my dream was shattered. I did not get to go! To this day that remains one of the biggest disappointments of my life.

Another special occasion, with a happier ending, was the trip to the Coca Cola bottling plant on Fourth Avenue between Centre and First Street West. The day we finished school in June, they gave all students a free coke and an oil-cloth wedge cap with "Coca Cola" written on it.

I first learned to skate on the Elbow River using "Bob skates." These consisted of an adjustable metal sole and sides to accommodate the size of your street shoes. There were two runners attached and leather straps that tightened around your ankles. The location on the river is where the park is located near Thirty-second Avenue and Elbow Drive.

There was a community rink across the street and down the alley about half a block from our home. I was on that rink morning, noon, and night during the week and all day Saturday and Sunday when I did not have chores or work. Homework ranked after work, chores, and hockey. All summer long, the goal nets were stored in Gord Dickson's back yard and we broke them out and played hockey with a tennis ball.

The School(s) of Hard Knocks

My first school in the Mission area was Cliff Bungalow Elementary where I attended grades five and six. The schoolyard was covered with cinders from the coal furnace so you did not dare fall, as the cinders would pierce your arms or legs. I played my first soccer against Elbow Park and other schools. I was a so-so goalkeeper. I was one of the first AMA Road Patrol monitors. We patrolled Fourth Street at Twenty-second Avenue West. We were issued white canvas gauntlets, a sash, and a whistle. A girl in our school accused me of hitting her over the head with my gauntlets and I received the strap (I was innocent). That was the first and last time I ever got the strap. I went to school with Bud

Selock, the Belzbergs, who lived across the street from the school on Twenty-second Avenue, Dave Laven, and many more.

Bud Selock and I thought we would be heroes and construct a boat to impress the girls. We "acquired" the wood from somewhere and were about to float it in the Elbow River, which was a block from my house, when we realized we needed a sealant to keep the hull afloat. The City of Calgary was repairing streets in our area. In those days large lumps of tar were melted by a wood fire in a trailer and administered to the cracks in the road. It was an efficient but mucky application. After completing a day's work the workmen parked the trailer and extra lumps of tar nearby. Bud and I were of the opinion that they would not miss a block of this material and we carried a piece of it home (it took the two of us to carry it as it must have weighed seventy-five pounds). We melted it and applied it to the hull of the boat in copious quantities—all seventy-five pounds. We carried the boat to the river with the help of a couple of girls, set it down on the water, and watched it sink. So much for the budding boat builders.

Near our home on Twenty-fifth Avenue and First Street lived the Simpson family: Sheila, Kathleen, and their mother; the Mac-Gregor family: Tom, Beth, and parents; the Dickson family: Gordon, Barbara, and parents; the Goronuk family; the Miller family: Marilyn, Barbara, and parents; the Wright family: Don, Doug, and parents; and the Simmons family: Dale and his mom. Dale eventually owned a multi-level producing oil company including pipelines and service companies. He died in about 2000. Also near us was the Jesee family: Fred, Glee, and parents. There were many more families with kids about my age living in an area of ten square blocks, such as the Barnetsons, Shields, Fayes, Moores, Taylors, Skeets, Campbells, Fyfes, and O'Learys, etc.

During 1942 I graduated from elementary school and started my junior high education at Rideau Park. This was an excellent school with kids attending from Parkhill, Parkview, Elbow Park,

Mount Royal, Mission, Roxborough, Rideau, and outlying areas. The principal was "Beaver" McLeod (called Beaver because of the big beaver coat he wore each winter). Other teachers were Dorothy Copp, Miss Nagler, Wallace Harper, Harry Hamilton, and P.N.R. Morrison. P.N.R. went on to be a City of Calgary Councillor and eventually owned a fairly successful oil company. He was a resolute CCF'er (currently named the NDP) and at one time had been a wrestler. P.N.R. could handle all the male students who considered themselves tough guys.

Rideau Park's Shop and Home Economic classes were held at an old bungalow school in Elbow Park about two blocks north of the Elbow Park Elementary School. Our route to the bungalow school necessitated crossing the swinging bridge over the Elbow River. Shop and Home Ec. classes were held once a week. I learned a most valuable lesson from our shop teacher, Mr. Elliot. Our first project was a leather-bound picture frame. The basic foundation was cardboard with leather front and leather lacing on the sides and top. While making the rear stand-up cardboard piece I did not score it properly and the cardboard had an out-of-place score mark. However, the front leather section was absolutely perfect. My mark was a B+ and I was extremely unhappy. I complained to Mr. Elliot and he explained: "Both front and back had to be perfect to get an 'A.'" When I protested, he replied calmly, "Bob, you can see both sides and you know the back is not perfect. Either side is as important as the other." I live that lesson to this day and it is one I have passed on to our children, grandchildren, employees, and any others who would listen.

Rideau was great for sports and I excelled at this level in hockey and fastball. We won the Calgary Junior High Seniors' hockey championship for the first time in the school's history. I am very proud of that accomplishment.

During spring and early summer our physical education, calisthenics, and free-for-all soccer were held out of doors. Wallace

Harper, our PT teacher, usually arrived late to conduct the class. A number of us would hide and catch a smoke in the bushes on the hill backing the school. When Harper arrived he ordered us to line up for our class. We ignored him and after warning us again, he would fetch the principal. At this point we would rush down and the lines were in perfect order when the principal arrived. Harper was very frustrated with us to say the least.

It was in junior high that I first learned to dance through a swing club started by some of the girls at the school. These were Anita Timmins, Barb Geddes, Sheila Jamieson, Pat Lewis (my first true love), Joyce Smith, Joyce Turner (my second true love), Barb Curlette, etc. Some of the fellows in this club were Graham Bennet, the Sinclair brothers, Don French, Dave Laven, etc. My mother arose from her sick bed to teach me how to dance. Thanks, Mom.

It was 1942 when my mom and dad first took me to the Shaganappi Golf Course when I was twelve. That was the beginning

Rideau Park Senior Champions, 1944–45. This was the first year the school won the city senior championship. *(Back row, L – R):* Trev Walters, Ken McIntyre, Bob Rintoul, Wallace Harper (coach), Bob Hebenton, Ron Oughton, Dick Douglas. *(Front row, L – R):* Omer Patrick, Phil Ross, Dave Laven, Dave Hebenton, Dale McReary, Jim Fyfe.

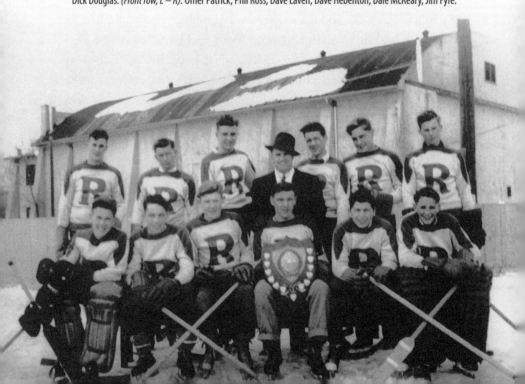

of a life-long passion for the game. In order to play at Shaganappi Golf Course my dad and I carried our clubs two blocks to the streetcar, then transferred to the Kilarney streetcar, which took us to Twenty-sixth Street and Seventeenth Avenue SW. From there we walked nearly a mile to the course, played a round, then reversed our steps. Shaganappi was a typical prairie course, with sand greens: a circle of sand coated with oil. To smooth the sand you pulled a piece of burlap attached to a steel handle.

We could not afford new golf balls and, besides, they were in short supply due to the war. We solved this problem by going on a golf ball hunt and then attaching a piece of thread to the ball with a straight pin and hanging the ball from our kitchen ceiling with another straight pin. We then used a fast-drying white paint to bring the balls to nearly new. The paint had a strong-smelling chemical that facilitated the drying process. I did not realize it at the time but this was like sniffing glue.

It was at Shaganappi that I had a very unusual and frightening experience. One afternoon while playing with another chap, lightning struck me. It happened on the fifteenth tee. My partner was on the tee (slightly raised with a rubber mat) ready to drive and I was behind and to his right, near a barbed wire fence. A fast-moving front blew in from the west and the next thing I knew, my golfing partner and I were lying flat on the ground and had lost our hearing. The bolt was obviously attracted to the barbed wire fence and had struck the earth about twelve inches from my foot, ran through me, and hit my partner in the forehead. A number of golfers had taken refuge from the impending storm in a wooden roofed, open-sided shelter. They thought we were dead! We regained our feet and hearing and upon inspection did not find any problems with our bodies or clothes. My mother and father were travelling west on Bowness Road, which runs below the escarpment, parallel to the golf course. They apparently wondered where the lightning bolt had hit. They found out when they arrived home a couple of days later.

In 1943, when I was thirteen, I was chosen for a peewee team sponsored by the Calgary Buffalo Athletic Association, which was bankrolled by the Calgary Brewery, owned by J.B. Cross. This was "A hockey" (at that time there were two classes of amateur hockey, "A" and "B"). I continued in this program through midget, juvenile, and junior.

The Calgary Buffalo Athletic Association provided sweaters, socks, pants, shoulder pads, shin pads, gloves, and a leather helmet for four teams per group totalling sixteen teams, including ice time for both games and practices at the only artificial rink in Calgary at that time, Victoria Park Arena.

What a thrill making that calibre of hockey. It was not unusual to practice at 5 A.M. Living so close to the Stampede Grounds, where the Victoria Arena was located, made walking to the arena in winter somewhat easier regardless of low temperatures. As we made our way to the arena, other team members joined our happy group. We threaded our hockey stick through the laced closure of our duffle bags full of equipment and slung the stick over our shoulders.

Each year two all-star teams were selected to compete to a "full house" of fifty-five hundred fans. A different division was selected each year. What excitement to play in front of so many people. I played with the Buffalos right through Junior B. Like every aspiring young hockey player, the NHL was my goal. I had the size, six feet two inches in height and 195 pounds, but I was not a strong enough skater. I was a stay-at-home defenceman who was pretty adept at throwing a hip check.

It was interesting to note that Doc Seaman's biography in 2008 mentioned him signing an "A" card. It was necessary to sign an "A" card before you could play for an "A" team. Once you signed the card you were the property of whichever NHL team issued that particular "A" card. Calgary was divided into districts and each district represented a different NHL team. If I made the NHL, I

Buffalo Midgets, 1946. Like many young Canadian boys, Bob dreamed of playing in the NHL. *(Back row, L – R):* Baldie Alteman, Herbie Hibbs, Pat McCaffery, Bob Rintoul, Peter Simon, Greg West, Alex Martin, Jim Jempson (coach). *(Front row, L – R):* Art Wells, Bill O'Leary, Johnnie Anderson, Jim (Mousie) Perkins, Jack O'Leary, Bob Dodderidge, Phil Ross.

was the property of the Detroit Red Wings unless I was traded.

I was fortunate to receive very few injuries during my short hockey career. I never wore shoulder pads, as they inhibited my movements. However, I did break my left hand and in my last year with the Buffalos I developed a severe hip pointer. To this day that hip bothers me and in my early twenties I had difficulty sitting for more than an hour.

In the 1990s I had a call from one of the executive of the current Buffalo Hockey Association requesting my assistance in organizing an alumni reunion. I jumped in with both feet and there were a large number who attended the reunion. We were able to include Mr. Jim Kerr, who was J.B. Cross's right-hand man. There were seven, including myself, attending from our original team.

The Alberta Sports Hall of Fame in Red Deer has accepted

pictures and equipment as well as historical information about the original Buffalo Athletic Association. It featured them in a display along with the Sutter family. These displays appear periodically in the Sports Hall. The Sports Hall of Fame also takes these old artifacts to schools, etc. The curator tells me that the young people cannot believe the flimsiness of our old hockey equipment.

I also played for the Sea Cadet hockey team when I joined Undaunted II in 1943. We had a pretty good hockey team and competed against any team that was available. Undaunted was located in a building designed for badminton (not used since the depression) on Rideau Road. The location eventually became television Channel 2 before the building was destroyed by fire.

My term of service with the Sea Cadets ended in 1945. We were taught seamanship by former and current Royal Canadian Navy hands and the training was excellent. I believe the training I received was instrumental in developing the discipline required for my future life skills. During the summer we encamped to the Sea Cadet facilities at Chestermere Lake. This was a two-week stint involving sailing, cutter (a large row boat) training, marching, etc. Once a month a group of us "old" Sea Cadets gets together for lunch.

Work Hard; Play Hard

My working career started at twelve years of age when I delivered groceries for a Chinese grocer on Fourth Street between Twenty-fourth and Twenty-third Avenue SW. The store is still there. There was no money in our family and I was beginning to feel out of place at school in my breeches and high stockings so I decided to do something about it. Fortunately, my father had bought me a CCM bike for my twelfth birthday. My family was pressed financially to give me such a wonderful birthday present. The Hay family provided bicycles to its children for their twelfth birthdays and apparently my dad's pride dictated that he do the same for his son. Lucky me.

I found a used metal carrier and mounted it on the front of the bike. These carriers were about thirty inches square with attachments to secure to your handlebars and down to the forks. The entire assembly swivelled as the front wheel turned. It was made of strap iron in a lattice configuration. A cardboard box of the proper size fit inside.

You could pack a lot of groceries in the carrier. I received twenty-five cents a load, which was pretty good money in those days. I worked the summer of 1942 and the winter of 1942–43. Sometimes during the winter the snow was so deep you pushed your bike and the load of groceries through the snowbanks. The deliveries were after school, on weekends, and during the summer holidays.

At Christmas the owner asked me to take home a carton of Dad's favourite cigarettes and a beautiful basket of fruit for Mom. He was a very kind gentleman. At the same time I worked cleaning store windows on Fourth Street and Twenty-fifth Avenue SW for McDermid Drugs and the candy store next door.

In the spring of 1943, I was fortunate to rustle up a job delivering meat for a butcher next to the Jenkins store between Twenty-fifth and Twenty-sixth Avenue on Fourth Street SW on Saturdays and during summer to customers in Elbow Park and Mount Royal. My farthest trip was to the Lougheed residence up the hill on Sifton Boulevard, the same Lougheed who was premier of Alberta. This job paid more than the job with the Chinese grocer and I was able to learn a fair amount about the meat business.

I also secured the *Herald* paper route for the Mission district from Charlie Money, who was moving to a better job. In short order I picked up a job at Safeway on Fourth Street between Twenty-fifth and Twenty-sixth Avenue SW, whose branch designation was No. 205. This was a Saturday and summer holiday job, so I farmed out the paper route for a 10 percent cut when I was unable to deliver papers.

I learned to stock shelves and prepare produce for sale. I was fortunate enough to score another paper route in the Roxborough district and I repeated the Mission route paper arrangement when the work at Safeway conflicted. I guess I have had an entrepreneurial spirit since I was a kid.

While attending junior high I held down a number of jobs and played hockey for three and sometimes four teams, as the community team would recruit me. I also had my schoolwork and I took on the added responsibility of selling advertising for our yearbook. Despite all of this, I managed to complete junior high with flying colours.

I planned to attend Western Canada High School, but my mother wanted me to attend the academic high school, which in those days was Central Collegiate Institute (the other high school students nicknamed us the "Calgary Collection of Idiots"). The school was located on the corner of Twelfth Avenue and Eighth Street SW. My first year at CCI was 1945–46 and in winter and summer I walked or rode my bike to and from school. Our home was on First Street and Twenty-fifth Avenue SW, a fair distance to bike or walk.

I did not do as well in high school as I had in junior high. I was intelligent enough but liked to goof off, and, truth be told, I really wanted to be at Western. I invariably hung out at Western during noon hours at the store across the street called the PB (it was referred to as The Pregnant Bunny). To this day there are a number of my high school contemporaries who graduated from Western and claim I went to their school rather than Central.

I played hockey for Central but the Physical Education teacher was not one of my favourite people. Therefore, I did not enjoy my high school hockey experiences. We travelled to Edmonton to play one of their high school teams, but I goofed off and visited some girls. I ran two mile in track but after running five miles every morning for eight months I cramped up at the track meet with two laps to go and did not complete the race.

Calgary Stampede parade, Sea Cadets, "Undaunted II," 1945. The cadets helped
Bob develop the discipline he would need in later years. Bob is second row, left.

My only enjoyments in high school were my jobs, Sea Cadets,
and hockey with the Buffalos. I was recruited to join the Delta Phi
Delta fraternity. These were great guys and we had a lot of fun
attending parties and dances. We had a frat hockey team and
played against the other fraternities. We usually wiped them as
Phil Ross, a Buffalo teammate of mine, was DPD and the other
frats did not have very good players. Phil and I would score six or
eight goals a game. Not much competition but good for the ego.

There were some redeeming features about CCI. Two of my
elective subjects were Bookkeeping I and II and Psychology I and II.
Our bookkeeping teacher was a fellow by the name of Mr. Assels-
tine who was reputed to have a steel plate in his head. I seemed to
collect a fair amount of detentions in his classes. His punishment

was an after-school session of adding twelve numbers per twelve columns across and twelve rows down. The results were proven in the lower right-hand column. You did not leave until you accomplished the feat. I became so talented at this exercise that I eventually could add numbers while viewing them upside down.

Mr. Brooks taught psychology and although I did not receive detentions in his class, I learned a great deal about what makes us humans tick. The knowledge I acquired helped me immensely in dealing with family, friends, acquaintances, and employees.

One of the stories still alive to this day wherever CCI students gather is when Minnie Maxwell, who taught English, dropped her upper plate on my desk. I was a bit of a class clown, so invariably was seated in the front row where teachers could keep an eye on me. Miss Maxwell was giving us a demonstrative dissertation of a Shakespeare play when out popped her teeth. I calmly picked them up and handed them back to her. This was an all male class and the mirth was instantaneous.

It took me many years to realize that I would have been much further ahead if I had ignored my dislike of certain teachers and got on with my education. I have attempted to dissuade my children and grandchildren from this unhealthy attitude. You make your mark in education; your teachers guide you.

The Sat-Teen Club was another organization I enjoyed. During the school term, the club met every Saturday at a different high school in Calgary (there were four: Central, Western, St. Mary's, and Crescent Heights) for a dance. Every so often they would bring in a big name band like Gene Krupa or Woody Herman. The club sponsored tours to different locations such as Banff, which we travelled to on old CPR colonizing cars. I was sixteen years old, and this was my first trip to Banff!

I eventually settled on the Safeway job and the two paper routes. I worked for Safeway until after I graduated from high school. When I was fifteen, I cut the paper routes loose. While at

Safeway I worked in nearly every branch in the city. I spent most of my time at #208, which was on Eighth Street and Sixteenth Avenue SW. This was very close to Central High School and was convenient to our home.

It was at #208 that I was given the daily job delivering the previous day's receipts (averaging twenty-five hundred dollars) to the Bank of Nova Scotia on the corner of Seventeenth Avenue and Fourteenth Street SW. I accomplished this errand on my bike. Try that today!

I continued my education in the retail grocery and meat business including cashier duties and eventually headed up their inventory control group. In those days most retail stores in Calgary closed on Wednesdays at noon. The inventory crew would start to work and, first of all, balanced the tills. Normally, each store cashier balanced his or her till daily at closing. We laboured until store opening on Thursday morning at which time we had tallied every item in the store and balanced the books. My top salary with Safeway, working five-and-one-half days a week from 7:30 A.M. to 6:30 P.M. was 150 dollars per month.

I was working at #201, the main store in Calgary, located at Eighth Avenue and Second Street East. My parents were on vacation, and I bought a meal ticket for five dollars at the Chinese cafe next door. For coffee and meals I used the alley door, through the kitchen and into the seating area. I usually took my lunch break around 1 P.M. and one day noticed a water tap above the oversized soup pot. I learned that when the soup stock lowered they turned the tap on and added water. After that I started taking lunch at 11:30 A.M.

The war was in full bloom while I was working at Safeway and rationing was in effect. I was able to trade ration coupons with customers for food items we were not using for ones that we needed.

The City of Calgary resurrected the scenic streetcar from the 1920s as a platform to sell Victory Bonds. The bonds sold at twenty-five cents a coupon and were available at schools as well as other locations.

When the end of the war was declared in August 1945, celebrations broke out all over Calgary. Schools were dismissed and business came to a halt. Citizens flooded the downtown, forming conga lines—sometimes blocks long—that wound their way through hotels and whichever stores remained open. To the best of my memory there were no acts of violence or damage.

Dave Storey built a Model T Ford from parts of two wrecks. I took one of my few vacations two weeks before the commencement of grade twelve. Dave and I along with another chap loaded the old T with sleeping bags and camping equipment. We travelled to Banff, Lake Louise, Golden, Invermere, and Cranbrook. In those days 90 percent of the roads were gravel and in a few spots very steep. If we could not climb a hill in a forward gear we drove the T in reverse, which was a favourite trick of Model T owners.

We refueled at Canal Flats, about halfway between Golden and Windermere in BC. As it was early evening we asked a group of fellows to recommend a good camping location. They gave us instructions and when we arrived we noted that it was in a bowl surrounded by thickets and the road from Canal Flats wound gently down a winding traverse above us. We set up camp and after dark, while sitting around the fire, we had a feeling we were being set up for a robbery. The chaps who gave us the instructions had looked a bit unsavoury.

Sid, the other member of our group, decided to sleep in the open and Dave and I started to kid him about mountain lions jumping off a nearby stump onto his sleeping bag. Our jesting did not bother him; he went to sleep. Then Dave and I heard rustling in the thickets near the hill. We decided to break camp and had a devil of a time waking Sid. On a whim we drove up the hill toward Canal Flats and there was their truck parked in the ditch. We did not have much money but who knows what might have happened had we still been camped there. We arrived back in Calgary two days late for the start of grade twelve. Pop Weir, the principal, was a very kindly old fellow but was not pleased with our tardiness.

When I was seventeen and in grade eleven, I started to go out with a girl on a steady basis. Her name was Marian Spence. She was a couple of grades behind me at Central High. She lived down by the Tivoli Theatre on Twentieth Avenue between Fourth and Fifth Street SW. As I stated earlier, it was a fairly long hike between my home and CCI. We walked the same route to school and one day I asked if I could walk with her and carry her books. A couple of days later I asked if I could take her to a show. One thing led to another and we started going steady.

In those days "going steady" meant that you came to a mutual agreement that neither party would date others. It was almost like being engaged before being married. Mr. and Mrs. Spence invited me to dinner on a weekly basis. They were an interesting couple, with three children: Marian and two boys. The boys were away, one was married and in the air force, the other in the army. The parents must have really liked me because they asked me to join them, even when they had other guests over for dinner. It took me a while to realize that they thought that since we were going steady we might be mates down the road. My dad caught on to it faster than I did.

The relationship ended amicably when I came home in September 1951 for my grandfather's funeral. By this time I was working in the oil patch and was away for a month or two at a time. We discussed our relationship and decided to go our separate ways. Over the years Nola and I have had the occasion to join Marian and her husband at a dinner organized by mutual friends. She was a lovely girl.

During the winter and spring of 1950, I was becoming less and less enchanted with Safeway. I had "graduated" from high school in the spring of 1948. Well, I did not actually graduate because I failed French and therefore was not eligible for a diploma. C'est la vie.

I had saved five hundred dollars for my first-year university tuition, board, and room. Without a diploma this was a no-go so I purchased a 1937 Ford four-door sedan with an eighty-five horsepower motor. In those days cars were equipped with mechanical brakes—far inferior to today's hydraulic brakes. Manifold heaters were also the norm and they transferred heat from the manifold through the firewall to the passenger compartment. Trying to keep warm on a cold winter day was a challenge.

The vehicle allowed me to travel around without resorting to my bike or walking. I drove it until the fall of 1950 and sold it for the same price I had paid for it. I learned a lot of lessons from that car. It was not mandatory to carry insurance in those days and my dad suggested I should insure the car. But because he never carried life insurance, I operated the car without insurance.

One time, after a visit with my cousin Howard, who was at the tuberculosis sanitarium across from Bowness, I had a serious accident. A semi-trailer pulled out in front of me at the bottom of the Centre Street Bridge, where there was a patch of black ice. I could not stop and was headed directly under the trailer. I swerved, hit a power pole, and become the owner of a wrecked car with no insurance. The upshot of all this was a picture of my wreck on the front page of the *Herald* and me out five hundred dollars for repairs. Fortunately, the only injury was to my ego. Having a dad who worked for the City of Calgary's transit system was a blessing. He was able to get me a deal on replacing the pole (I had to pay for it as well). Although I pressed the trucking company for damages, they would not listen to a punk kid.

I was always tall for my age and at age fifteen could pass for twenty. This allowed me to drink beer in the different beer parlours in Calgary. However, I had to be cautious that I did not frequent the same bars as Father, who after a hard day's work would scoff a beer or two. Although I am not proud of myself, I started to smoke when I was twelve years old. My father smoked, as did a large percentage of adults. My mother did not smoke and had her first taste of an alcoholic beverage at the age of thirty-eight. At that time the negative health problems were largely unknown, particularly cancer. I eventually quit smoking in 1964 at age thirty-four.

I was now off to join the working world. My experiences with Safeway, the paper routes, and being raised in a union home, as well as the Sea Cadet training allowed me to tackle the next phase of my life with confidence. ∎

Part Two

The Oil Patch

During the spring of 1950 my mother was on the lookout for a job for me in Alberta's fledgling oil business. She played bridge with a group of women whose husbands were, among other things, salesmen at Maclin Motors. Maclin was selling vehicles to exploration crews. One of the ladies gave Mother the names of companies and people to contact. I called two or three before I scored with Northwest Seismic Surveys Limited.

I started with Northwest in April of 1950 at 150 dollars a month. If you lasted one month (and many did not) you received a raise to 165 dollars. I was immediately sent to Edmonton to work on new cables that cost eighteen-hundred dollars each—a lot of money in those days. We worked out of the Carter Oil laboratory (a division of Imperial Oil), located in the west end of Edmonton at about 140th Street. The last half mile into the lab was accessible by a bush trail. Our job was to cut the cable at the appropriate location and select the proper wires so the geophones could be attached to the inner wires of the cable in the proper order.

We returned to Calgary and assembled a crew and equipment and travelled to Bashaw, a medium-sized town in central Alberta.

We roomed and ate at the local hotel until we could find a "rooming house." I started as a jug hustler and in short order was promoted to the reel man's job on the back of the reel truck. Don, the reel truck driver, and I found a room at Mr. and Mrs. Hume's, proprietors of the lumberyard in Bashaw. We continued to eat our meals at the hotel, purchasing a five-dollar meal card divided into twenty-five-cent increment punches. You could purchase a steak dinner, soup, and dessert for $1.50. Breakfast would run you about twenty-five cents, and you received a discount for buying the meal card.

We worked hard, 7 A.M. to 6 P.M. with a 15-minute lunch break. We also played hard, often partying all night with no sleep. I could get away with three sleepless nights but would be in bed on the fourth evening by 6:30 P.M. We were required to work 220 hours per month. The time was accomplished either in straight days or with weekends off. The observer usually determined this choice, as recording hours were the official measure of a complete month's work, and you can be assured that if the observer had a personal commitment we worked straight through. Inclement weather cut into time worked and on occasion we had no time off for a couple of months.

While I was away working on seismic crews I insisted my mother and father use the car (Father was reluctant to do so), as they did not own one. The first car my parents purchased in 1951 was a 1940 Plymouth. I sold the Ford to an older gentleman who did not have a driver's licence. He asked me to drive him around a few blocks and then bought the car. He took me to his room and dug under the mattress and gave me five one-hundred-dollar bills. I had never seen a hundred-dollar bill.

Seismic Exploration

For the uninitiated, I will explain how the early shot hole seismic exploration worked. When an oil company suspected the presence of an oil-bearing structure it contracted a seismic company to carry out a survey before purchasing a lease from the provincial government and drilling a well. The seismic crew moved to the area that was to be surveyed.

The early seismic shot hole operation consisted of a surveyor and rod man flagging in shot hole points every one-quarter mile and geophone intervals every 110 feet, then surveying and noting the elevation and horizontal readings. The dominion government provided vertical elevations at all railway crossings and benchmarks on the borders of the provinces. The dominion government also provided horizontal locations on topographical maps.

The portable drills would drill a shot hole (anywhere from sixty to three hundred feet in depth) every quarter mile. The reel-truck crew laid out the cable, a quarter mile on either side of the shot hole. A large power-driven drum on the back of a pickup or the recording vehicle facilitated the laying out and picking up of the cable. The reel operator also dropped off the geophones at the required intervals. The jug hustlers attached the geophones to the cable, one every 110 feet and two every fifty-five feet. In other words, three geophones per trace for a total of twenty-four traces.

The shooter loaded the shot hole to the required depth with up to one hundred pounds of explosives and detonated the charge. The shock waves generated by the explosives' energy were picked up by the geophones and connected and transmitted by the cable to the instrument truck, where the results were recorded on a strip of paper to be analyzed by geophysical interpreters. After the first hole was shot and recorded, the back cable and geophones were picked up and laid beyond the next shot hole and so on.

A seismic field crew consisted of a party chief usually doubling as the interpreter. If there was no interpretation in the field then a party manager ran the crew. Other crew members consisted of a surveyor and rod man, an operator, a junior operator, a shooter and helper, two to four jug hustlers, a reel truck driver and reel man, two to three portable rotary drills complete with a driller, roughneck, and water truck driver.

The crew was expected to maintain the equipment, trucks, instruments, cable, and geophones in tip-top shape—this maintenance was on your own time whether at night or on "weather days." However, we seemed to find plenty of time to drink copious amounts of beer at the hotel bar. The bar opened at 11 A.M., closed for an hour from 6 to 7 P.M. (an Alberta Government regulation to supposedly force married fathers to go home and have dinner with their families), and reopened until closing at 11 P.M. You could buy bottled beer by the case before the bar closed. The price of bottled beer purchased at the bar was considerably more than from an official Alberta Liquor Board store. Unfortunately, these stores were located in larger centres or cities. You were required to purchase a yearly permit to buy booze at the provincial government stores.

When we arrived back in town for the day and if no maintenance work was required on the equipment, we headed for the hotel bar and started inhaling beer at ten cents a glass until the bar closed for supper. It was an unwritten law that as many glasses of beer as possible cover the table. When the bar was closed for dinner, the beer that remained was left for our return at 7 P.M. The drinking continued until the 11 P.M. closing when four or five cases were purchased to continue the party.

The "locals" and farmers also frequented the bar and it did not take long to make their acquaintance. On occasion there was the odd sponger who never bought a round. One of our drillers solved the problem by hanging the sponger by his bib overalls on a coat hook. The hook was normally high enough that the offending sponger could not touch the floor. The other trick was to pin the sponger's shirtsleeve to his pants as he was leaning his elbow on his thigh. When he attempted to rise he fell face down on the floor. These two remedies usually solved the problem of spongers.

In the early 1950s, oil exploration crews were a new experience to these small-town residents and we were greeted with open arms. The women at the rooming houses enjoyed discussing the nocturnal habits of their oil patch renters. However, after Nola and I were married we noticed that the price of goods went up considerably after an oil crew came to town.

We stayed in Bashaw from the end of April to mid June. As a matter of fact, I was in Bashaw for my twenty-first birthday, and my parents sent me a very expensive Hamilton wristwatch. My mother's birthday was on May 15 and I had forgotten to contact her or to at least send her a card. I found out later that my dad was furious regarding my thoughtlessness. I would not have received my birthday present if Mother had not prevailed.

Peace River Country

Our next contract with Imperial was near High Level, Alberta. Don Carter and I with the reel truck and Jim Thompson and Scotty, his helper, with the shooting truck and explosives storage trailer were the first to hit the road. The distance was over five hundred miles of mostly gravel roads. After a party night in Edmonton, Carter was anxious to get there so we travelled night and day spelling each other off at the wheel. We were on hard top roads to Westlock and from there it was gravel or dirt.

We travelled Highway 44 through Fawcett to Hondo and then Highway 2 to Smith and Slave Lake where we fuelled up. Slave Lake in the summer of 1950 consisted of a hotel, store, and gas station. About five miles west of Slave Lake the vehicle quit. We eventually determined that the fuel tank contained water. In those days, independent truckers would travel the northern highways and sell fuel to the independent service stations, which paid cash for the load. Sometimes a crooked truck driver would sell a load of water and pocket a handsome profit. The retail price of gas in the north was fifty cents a gallon whereas the price in Calgary was eighteen cents a gallon.

We got a tow back to town and had the tank removed, thoroughly drained, and dried. After a fill of "good" gas, we were back on the road again. We had lost about five or six hours so Carter was really in a stew to get to our destination as fast as possible.

From Slave Lake we continued on Highway 2 to High Prairie, the French communities of McLennan and Donnelly, and then on to Peace River. What a sight the Valley of the Peace was from the top of the Peace River hill. The confluence of the Smoky and the Peace rivers to the west was outstanding and I decided then and there that when I died, my ashes would be spread over that point in the rivers.

We wove our way downhill into Peace River and proceeded up the main street. A group of aboriginal girls was leaning out a hotel window beckoning us up for whatever (you have to use your imagination at this point). From Peace River town we crossed the Peace and wound our way up the other side of the valley. We continued to Grimshaw, which was the end of rail, and travelled north on Mackenzie Highway 35. While stopping for a few minutes we encountered a truck driver who had a bear cub, which he had recently captured, on a leash.

Prior to 1949 all goods to the north of Grimshaw were transported during the winter on sleds pulled by Caterpillar tractors. After leaving land at Hay River, NWT, they continued over the Great Slave Lake ice to Yellowknife.

The construction of the Mackenzie Highway immediately followed the construction of the Alaska Highway and men and equipment moved from that project to Grimshaw. The finished road was eighteen feet wide with a gravel surface. In many places it was constructed over muskeg (muskeg is a wet area of decayed vegetation and trees that may be thousands of years old and in many cases bottomless). During spring, summer, and fall these sections would on occasion swallow a semi-trailer truck. The gravel covering the road was known as "Texas Pea Gravel." Some single stones were as big as two fists. Approximately one hundred miles north of Manning a truck passing in the opposite direction threw a rock that had been wedged between the dual wheels. The rock smashed our windshield and plopped on the seat between us.

The contract ended in October, at which time we returned to civilization and had the windshield repaired. Later we were issued Dodge Power Wagons to better cope with the wet trails.

The Mackenzie Highway and River were named after William Mackenzie, a world-

Bob began his career in the oil industry with Northwest Seismic Surveys Ltd. Bob and Al Anderson *(right)* are standing by Northwest's Dodge Power Wagon near Fort Vermilion, 1950.

renowned explorer who traversed the river system from Fort Churchill bordering the Hudson Bay to the Arctic Ocean and also through the Rockies to the Pacific Ocean.

We passed through many small communities with names such as Notikewin, Hotchkiss, Keg River, and Paddle Prairie. We eventually came to a point on the highway where a crudely written sign advised us that we were at the High Level–Fort Vermilion junction. There was absolutely nothing at High Level except clouds of mosquitoes and black flies that took a chunk out of anything walking, crawling, or standing still. We were in a quandary—where was the camp? At that point this huge fellow walked out of the bush with a cloud of mosquitoes and black flies circling his fedora-hatted head. We inquired as to the whereabouts of a camp either locally or at Fort Vermilion. He answered in the negative. We quickly understood why the bugs were leaving him alone. He positively reeked of garlic!

We decided to head for Fort Vermilion, some forty-eight miles to the east. The trail was gravel free, and as June is the rainy month in this area the road was a sea of mud. We used the winch on the shooting truck, continually, and finally reached the northern shore of the mighty Peace River at 4 A.M. We had departed High Level the previous day at 10 A.M. We noticed a small ramp and agreed that it must be the ferry docking area. It was nearing the longest day of the year at this far northern location so the sun had barely set in the west.

The ferry showed up at 9 A.M. powered by a Model A Ford motor, which by a series of gears pulled the ferry along on an overhead cable. The current was very fast and it took us three-quarters of an hour to cross the river, which is about a mile wide at this point. On arriving at Fort Vermilion we were told that there was no oil camp in that area. We stopped at the ferry operator's house, which also provided meals to the public, and dined on homemade bread, moose meat, potatoes, peas, and fresh raspberries. We were surprised that the peas, potatoes, and raspberries were fresh but soon realized that the growing day this far north is about twenty-two hours.

The ferry returned us to the north side of the river and we winched our way the forty-eight miles back to High Level. Our sleep for a couple of days consisted of catnaps. We drove about five miles north on the Mackenzie Highway from High Level and there was our camp!

George Schultis, Imperial Oil manager for the Peace River district, was there to greet us. His comment: "You're early. You're not supposed to be here until tomorrow." The four of us were tired, grubby, and hungry, and his greeting was not what we expected. The current population of High Level is close to forty-two hundred. Our original campsite is the current sports grounds.

We eventually shared our campsite with a Big Rig crew (these crews dug the deep wells that penetrated the oil-bearing formations and hopefully hit oil or gas). In the early days, the service companies did not drive back and forth to Big Rig locations as needed but remained in camp until the well was completed. There were approximately one hundred men in camp, which afforded us the opportunity to man six fastball teams. We set up two ball diamonds and played games every evening and, with nearly twenty hours of daylight, we sometimes played double headers. It was obviously of interest to the local black bears as they used to come to camp to raid our garbage dump and often sat on a bank about five hundred feet away and watched us play ball.

One day I found a chicken hawk with a broken wing. I took him to camp and assisted him in recovering the use of his wing. The hawk was with us for about a month. He would sit on the end of my bunk at night and never left a mess. During the early part of his stay I would feed him pieces of meat and when he started to fly short distances I would take him outside, give a whistle, and the cook would come out of the cook shack with a piece of meat and the hawk would fly over for his meal. Eventually, I took him to an open area and threw him into the air where he started to fly. He circled higher and higher, then came back for one swoop over my head and disappeared.

Imperial Oil had a horse exploration crew working the Caribou Mountains northwest of High Level. The crew consisted of Imperial employees, most of whom were graduate geophysicists except for the cook, his crew, and the horse wranglers. They dug their shot holes with hand augers, which allowed them to be very mobile. We were their contact with the outside world. Communication at that time consisted of two-way radios of very low wattage. We were required to relay messages from the crew to headquarters in Peace River. Their supplies were air-dropped. One of the crew members was a good friend of mine, and I met him in Peace River after they closed the crew down following four months in the bush. They were a pretty ragged bunch of men. Alec Mair said to me, "Here I am, an honours geophysical graduate, and I'm riding horses. I'm going to get my doctorate." He did, and Imperial shipped him to South America.

The available water from local muskeg-drained and silted rivers was very uninviting, however potable. We were fortunate to find a spring about two miles north of camp that poured forth a stream of clear pure water. We had to transport it with milk cans and then transfer it into large war-surplus fuel tanks. The tanks were salvaged from Lancaster bombers, the largest plane operated during World War II. The tanks were approximately two feet high by fifteen feet long and ten feet wide. Our camp attendant could move these empty tanks at will. He was extremely strong and could throw a full forty-five-gallon barrel of fuel over the stock rack of a truck. He enjoyed taking on three of us young guys at wrestling and would "whup our asses." He shot a moose about two miles from camp, halved it, and dragged the two halves to camp.

After a month or so we moved to Rocky Lane on the north

side, adjacent to the Peace River where the ferry crossed to Fort Vermilion. There were a few farms carved out of the bush mostly by Ukrainians who had moved from Saskatchewan during the depression. They encountered rich soil and nearly twenty-four-hour sunlight, which afforded fast maturity for crops. But late and early frosts meant the overall season was short. There were two camps of farmers, one north of the Boyer River and the other adjacent to our camp at Rocky Lane. Each settlement included a school, teacherage, and community hall. On alternate Fridays the teachers would host a dance and the locals and us would have a great old hoedown. I was seeing the teacher at Rocky Lane, which was a very convenient arrangement.

The locals made moonshine and at a dance during the harvest season they introduced our crew to the "niceties" of this potent drink. We had one very inebriated crew. One member climbed a tree in front of the community hall and fell, "passed out."

During our stay at Rocky Lane we visited Fort Vermilion to play ball against the Canadian Agriculture Experimental Station employees and after the game they held a dance. We did not lack for entertainment.

After we completed our necessary hours for the month we would pile into a couple of Power Wagons and head to Calgary. It was a long trip and we took turns riding in the cab. The drillers took their personal vehicle(s), as they lived near Camrose. They returned to camp a few days before the rest of the crew to repair the drills and have them operational for the next month's work. We arrived at camp to find fifty to seventy-five Natives surrounding the camp. The reason? The drillers had set up a record player and attached a speaker and music was pouring out at high volume.

While at that campsite we witnessed the granddaddy of forest fires. The fire travelled mainly west to east, north of Paddle Prairie, and crossed the Mackenzie Highway. The fire was so severe and the smoke so voluminous that it was reputed to have travelled on the prevailing winds to Great Britain and reduced the visibility at the London airport. Planes were unable to land or take off for days. The fallen ash at our camp and surrounding area was about eight inches deep and we did not see the sun for two weeks. The nights were pitch black and you could barely see your hand held up to your face. We continued to work, but the RCMP closed the Mackenzie Highway to the town of Peace River. The only people allowed to travel were those responsible for providing our supplies. We were forbidden to smoke and turned to chewing tobacco. It was most unpleasant staying in the small four-man bunkhouses, as syrup cans were used to expectorate the residue tobacco and we were not very good shots.

Roughing It

We completed our contract in the High Level area in October and moved the equipment to Calgary. Our next contract was based out of Camrose, Alberta. We purchased our winter clothes from a war surplus store on the south side of Eighth Avenue between First and Second Street East. The woolen long underwear had been manufactured by Stanfield's. Our pants and shirts were heavy-duty army issue khaki. The parkas were complete with a closely woven outer shell, hood, and the inner liner was real lamb's wool. Our boots were air force issue with lamb's wool inner liners. We wore sheep-lined air force helmets complete with attached earmuffs and leather mitts with wool liners.

We moved the crew to Hondo, Alberta, on Highway 44 just south of the northern junction of Highways 44 and 2. The village of Smith was ten miles north of Hondo. Imperial Oil had expended its supply of bush camps so we were forced to use one of the Northern Alberta Railroad's maintenance railway camps consisting of three sleeper cars, each of which had bunks for twelve men. The inside of the sleepers contained six double metal bunks, a coal space stove, a small shelf for a water basin, and a forty-five-gallon drum to hold water. We removed additional bunks (originally sixteen) to set up desks for office work. We equipped the barrel of water with a floating block of wood to attempt to stop the "slopping" of water when we were shunted in the middle of the night to Smith where there was a side track to allow through trains to pass. The camp also consisted of a cook car and water tender.

The interiors of the cars were absolutely filthy, so we fumigated and whitewashed ceilings and walls and purchased new mattresses and blankets. The floors were wooden on top of the metal frame and a few of us placed our dress shoes under the bunks where they froze to the wood, requiring careful removal. The cars were not insulated. If a blanket touched the wall it was frozen in place the next morning.

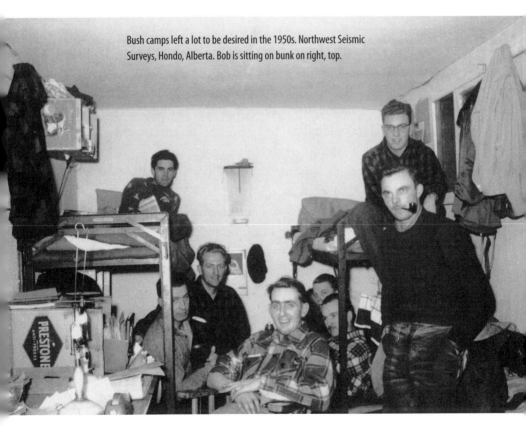

Bush camps left a lot to be desired in the 1950s. Northwest Seismic Surveys, Hondo, Alberta. Bob is sitting on bunk on right, top.

The NAR provided a camp cook whose experience was limited to lake fish, hamburger, and salt pork along with vegetables. We introduced him to the fine art of cooking steaks, ice cream, and frozen strawberries including the best of supplies we could muster. It certainly was an experience. We occasionally opened up our lunch bags to find "green ham" sandwiches because the cook did not feel it necessary to refrigerate the ham. Our only water supply was from the railway water tender and there was always a film of ice on top. As a result, every morning the smallest of our crew would be lowered through the manhole on the top of the tank to break the ice and "pail" the day's supply of water back through the manhole.

The winter of 1950–51 was particularly cold. During one stretch, the temperature never rose above minus 50 degrees Fahrenheit for two weeks. Our drills had nine-inch pump cylinders and if not moved from shot point to shot point with the mast up and all systems operating, the components froze.

Prior to World War II, our party manager, Gene Cook, was engaged as a schoolteacher. Gene used his DVA credits to attend university and graduated as an engineer. Due to a lack of entertainment and the additional time available, Gene offered me the opportunity to review his engineering books with his tutelage. I jumped at the opportunity.

During the war he was a navigator in the Pathfinder Flying Division of the RCAF, which charted the routes the Allied bombers used to bomb Germany. It was a very dangerous occupation, as the planes had no weapons. That winter Gene quietly boarded the train to Edmonton. A few days later the *Edmonton Journal* arrived in camp and lo and behold Gene had been decorated with the Distinguished Flying Cross for his war accomplishments. Gene was a very humble man.

Al Anderson was the surveyor and I was his rod man. We explored the area between Highway 44 and Highway 2, which was very rugged country. One day we attempted to turn around on a cutline and the fan pierced the radiator. The survey crew was always ahead of the rest of the crew and in many instances was in another part of the prospect. We were stranded in minus 30 degree weather with a heavy snowfall. Eventually, around 10 P.M., Lloyd McNee and Luther Lakevold found us and towed the truck back to camp. On their way out to find us they helped a motorist out of the ditch. The fellow sped off without thanking them. The boys caught up to the car and pushed it back into the ditch.

We had a rather unusual group of men on that crew. One fellow lay in bed and cried. The following spring we realized the reason for his sadness, which I will relate in a few paragraphs. Charlie was the jug truck driver and he felt he was God's gift to women (although none of us could figure out why). He invariably found an excuse to go to Smith, some ten miles up the highway, every evening to visit at the garage and have some imaginary item on his truck repaired. It did not take us long to find out that the garage owner had a very attractive wife and Charlie would regale us with stories about how much she liked him. We'd had about enough of this foolishness and one evening when he returned after we were in bed, he fell through the springs on his lower bunk. We had separated them. Another time we found a deer, more dead than alive. We felt we could bring it back to life with some TLC. One evening while Charlie was on one of his trips to Smith, we placed the very languid deer in his bunk. Needless to say there was quite a commotion when he attempted to crawl into bed.

The stoves in the railcars were space heaters that burned coal. We hired a local fellow to keep the fires burning during the night. He had a very attractive wife and some of the boys would visit her while he was busy keeping the fires going.

We worked out of this camp until spring breakup and headed back to Camrose. This was a great place to work during the summer. Our office was located in the basement of the Alice Hotel with one little window for ventilation. We cleaned our survey instrument with carbon tetrachloride (the fluid they used in most portable fire extinguishers in those days—1951), which we subsequently discovered was very carcinogenic.

Now back to the "crier" at Hondo. Lloyd and I were attracted to a very good-looking young woman who lived with her mother in a small trailer behind the Esso service station. It turned out that the "crier" had impregnated the daughter a year or so ago, and the mother said, "You have to marry me," which he did. No wonder he cried at night.

Boys Will Be Boys

We had many good times in Camrose and there were many parties and dances. It was after one of these dances that I received a very debilitating injury. However, it did not manifest itself for nearly fifteen years. On the way home from a dance at Buffalo Lake a crew member ran into the back of my truck while I was stopped at a stop sign, whipping my head back. Finally, in 1966, I had to have three vertebrae in my neck fused.

One night Charlie (you remember our lady's man) decided to purchase a mickey (thirteen-ounce bottle) of Crème de Menthe to impress the girls at a party. An employee who had recently been mustered out of the army joined our crew. He considered himself a macho man and viewed the remainder of the crew as a bunch of punks. He constantly boasted about his prowess as a drinker. At the party he knocked back the entire bottle of Creme de Menthe in one breath and disappeared. He stayed in the same rooming house as a number of us. It was apparent the following morning that his bed had not been used. Around noon that day I decided to search out his whereabouts. I found him passed out in a field near the home where the party had been held. Fortunately, he was alive.

During July 1951 I received orders from head office in Calgary to catch a bus to Biggar, Saskatchewan, where I was to be the

surveyor on a slim-hole crew. I would use the electric log operator as my rod man and assist him in the E-logging.

Slim holing, also defined as structure test drilling, is the simple process of drilling up to two thousand-foot holes to search for
oil- or gas-bearing structures. The drill rig is mounted on a five-

Slim hole drilling, Medicine Hat,
1951. This process consisted of
drilling up to two thousand-foot
holes in the search for oil and gas.

ton truck chassis. Rather than daylight hours, as in seismic exploration, the rig operates around the clock in three shifts. Each shift includes a driller and two helpers. The holes are drilled to the required depth, and E-logged, and in some instances core samples are retrieved. These rigs can also be used for shallow gas production.

E-logging is basically the gathering of sub-surface information to facilitate the mapping of potential oil- or gas-bearing structures. An electronic-powered probe is lowered into the hole by a cable and emits electronic pulses to an instrument that interprets and prints the information on paper (known as the log). The log readings differentiate the densities of the structures. That information is used to produce a sub-surface contour map.

Biggar was a great party town, being a division point for the CNR. There was also a Big Rig crew in town at the time. The CNR operated an all-night cafe at the station to facilitate meals for the railway shifts. It was a great place for shift workers to hang out and catch a fast meal either after or before a shift. The waitresses and other girls who frequented the café were an added attraction.

The tool push, E-log operator, and I roomed at the hotel. The rest of the crew stayed at various rooming and room-and-board locations. I was fortunate to meet a young lady whose dad took a shine to me and offered his car whenever his daughter and I were on a date. He was the local barber (there is a follow up story, which I will detail later).

The day shift planted about seventeen hundred feet of drill pipe in the hole. It would not move, either up or down, or circulate. The pipe in the hole was worth around five thousand dollars (in 1951 dollars). We tried everything to free the pipe, with no success. The Big Rig crew offered their help; still no success. When we had an unsolvable problem, we usually retired to the beer parlour and contacted the Calgary office. Our drilling superintendent agreed to come to Biggar. In the meantime he ordered us to have some dynamite and detonators delivered. If all else failed we would "dynamite" the string of pipe.

I ordered the dynamite and detonators and retired to the beer parlour. When the order arrived the truck driver asked us where our explosives magazine was located. We replied, "What explosives magazine?" Consequently, in a rather inebriated condition, I accompanied the load to the local graveyard and hid the explosives and detonators behind a headstone. When the superintendent arrived he asked me to retrieve the explosives. I could not remember behind which headstone I had hidden the explosives. We finally located the explosives and detonators but were unable to remove the pipe. To the best of my knowledge the pipe remains planted in the ground on a section road near Biggar.

In September we were issued marching orders for the Alberta/Saskatchewan border near Medicine Hat. We were instructed to stay at Irvine, which is located on the TransCanada Highway about thirty miles east of Medicine Hat. When we arrived in Irvine, there was no place for the crew to stay so after a phone call to the president, he agreed to allow the crew to headquarter out of Medicine Hat.

My grandfather Burke died in early September and I travelled to Calgary for his funeral. As an aside, Highway 1 (now called the TransCanada Highway) was not completed from Suffield to Tilley. It was a cattle trail so you were obliged to use Highway 3 to Lethbridge and Highway 2 through Fort Macleod to Calgary. While in Calgary Marian Spence and I agreed to discontinue our steady relationship.

Lloyd, Gooch, and I rented a motel just out of town on the way to Lethbridge. We cooked some of our own meals and ate the remainder at a restaurant. There were fourteen single females to each single male. What a deal! We would park our car or truck on the main drag on Saturdays and the girls would pick us up.

Smitten

It was party, party, and more partying. I dated a girl by the name of Marion Byers for a while and after a big party one Saturday night we were cruising around on Sunday and decided to pick up Marion's cousin. While waiting for her, a very pretty girl crossed the street. On inquiring who she was, I was told her name was Nola Cooper.

At that time I was taking inventory of my life. I did not like what I saw and looking into the future was even more depressing. I was working on an oil exploration crew, drinking and partying excessively, and spending whatever money I had left between paydays purchasing unnecessary clothes. This was a mirror of my two buddies on the crew, Lloyd and Ron. I decided a change in direction was necessary and the change arrived when I officially met Nola Cooper.

About a week later I was in the field for two days attempting to solve an E-logging problem. The crew delivered my meals at shift change. I finally solved the problem and drove to our motel, dirty, tired, and ready for a shower and some sleep. Lloyd arrived and asked me to accompany him, his girlfriend, her mother, and two other girls to Redcliff for a beer. I said, "No, thank you," but he insisted and I complied but warned him I was not changing my clothes or showering. Fine by him. It was Tuesday, September 25 (this date is an important factor down the road). When we went to the car there were two young ladies sitting in the back seat. One of them, who was obviously very aggressive, shoved the other over to the door and said sweetly, "Sit here." I said, "No, if I'm going to sit in the back seat with two lovely girls, I'm going to sit in the middle." The other girl was Nola Cooper.

We drank a few beers at the Redcliff bar (Nola would not go to a bar in Medicine Hat because she worked at Household Finance and felt it was inappropriate for her to be seen in a local bar). At Nola's request we returned to Medicine Hat reasonably

Bob standing by one of the few trees near
Medicine Hat, Alberta, 1951.

early. Apparently, Nola had recently endured a couple of unhappy experiences with men and was not very receptive to *this man*. She insisted on being taken home first. When we arrived at her rooming house, I walked her to the door and asked her, "May I give you a ring (telephone) sometime?" She said, "Oh, I guess so, if you want" (in a very unpleasant tone of voice) and gave me her telephone number.

Subsequently, Nola has explained that as she was walking to her upper floor room, she hesitated on the landing before the stairs and said to herself: "Sure he's going to give me a *ring*; this is the man I'm going to marry," and the thought immediately left her mind. The next day, Wednesday, I phoned her for a date and as she was catching up on her housework, she invited me over so we could get to know each other. The following day, Thursday, I asked her to a movie and Ron Elder, one of my buddies on the crew, joined us.

On Friday, September 28, three days after I met Nola, she invited me for a home-cooked meal. I felt this was the woman with whom I wanted to spend the rest of my life. I proposed and she accepted. I told her that we should make our wedding plans now as the crew often moved on very short notice and I might never return!

We planned the wedding for October 27. However, when we talked to my parents they had been invited to the State dinner for Princess Elizabeth and Prince Philip (due to Dad's position as a senior union representative in Alberta). Nola said, "I cannot ask your mother to choose between the State dinner and the marriage of her only child." So we moved the wedding date to November 3.

We were married in the choir loft of St. Barnabas church in Medicine Hat. We picked the "Hat" as it was relatively equidistant between Moose Jaw (where Nola's mom and brother lived) and Calgary, where my parents lived. Our parents did not meet until the wedding day. Other than our parents, Nola's mom's friend, Hugh, and Nola's brother, the wedding guests included Lloyd McNee, his girlfriend Betty, and the Burns—Al and June—who were best man and maid of honour.

Our wedding dinner was held at the Glass House (long since gone), which overlooked the South Saskatchewan River. We had a fried chicken dinner and the bill for eleven people was twenty-five dollars (paid by the bride). It snowed that evening and the snow was still on the ground the next morning. My premonition was correct—the crew moved out of Medicine Hat the next day.

Mom and Dad drove Nola and me back to Calgary and we stopped in Taber to see my aunt Grace, uncle Howard, and cousin Blair. Blair raised English Cocker Spaniel dogs, which were of excellent lineage. I made up my mind that we would eventually own one of these dogs.

Nola and I stayed at my folks' place in Calgary and I went into work on November 6. About November 15 the crew moved to Oyen, Alberta, near the Alberta/Saskatchewan border on Highway 9. We loaded a wooden barrel with our worldly goods, along with suitcases, in the back of a half-ton pickup and away we went. Oyen is and was a very small town. They had one water well in town and living accommodation was at a premium, particularly for a married couple. We could not afford to stay in the hotel at three dollars per night.

Finally, after much discussion with the local folk, we were advised to approach Mrs. Trewin. The locals thought she might like to rent out part of her home as her husband had died recently and she was living in the kitchen. The remainder of the home was closed up to save coal during the winter. She agreed to allow us to open up the rest of the house and said we could have the front bedroom. She moved into the other bedroom and we had the use of the front room and kitchen. The rent was $42.50 per month, which was much better than ninety dollars per month at the hotel. There was no running water and the toilet was outdoors.

Nola had to prepare our meals when Mrs. Trewin was not home, and we more or less stayed out of each other's way. One night we heard Mrs. Trewin talking to someone in her room (the walls were very thin). After a few nights we determined that she was carrying on a three-way conversation with her dead husband and equally dead dog.

When Nola wanted to wash clothes, she received permission from Mrs. Trewin to haul the electric washing machine in from the back porch and thaw it out. It was an electric washer with wringers. The clothes were hung outside to dry. It did not take long for them to freeze solid but when carried into the house the fresh smell was a treat. It was extremely cold that winter and the temperature at one stage never rose above minus 50 degrees Fahrenheit for two weeks. The outhouse seat developed a ring of frost that made sitting rather uncomfortable.

On a daily basis, I drove the one-ton truck to the town well and filled four one-gallon pails with water. When I arrived at the house I was lucky to have four half-filled pails of water complete with a skating rink on the floor of the truck box. Eggs were very difficult to come by and as Nola was making lunches for me every day, she wanted to provide some variation. Finally, a gal we met in the bar suggested Nola visit the grocery store during lunch hour when the owner was away and she would sell her some eggs. Nola came home with three eggs wrapped in Kleenex—the cost, three dollars!

The rig was mounted on a five-ton Ford truck powered by the largest gasoline motor available in those days. The crankshaft failed and the interior of the garage in Oyen could not accommodate the truck and mast. We covered the cab and motor of the truck with a tarpaulin and heated the covered area with propane heaters. We ordered a motor block to be delivered to Seismic Service Supply in Calgary. Two of us drove to Calgary to pick up the motor block. When we returned to Oyen, the temperature had dropped to minus 58 degrees Fahrenheit. We exchanged motor blocks within hours under extremely difficult conditions.

The exploration area consigned to us was in the vicinity of the Red Deer River. It was twenty-five miles from one ranch home to another. We drilled a hole near one of these ranches on the river breaks and encountered gas at a depth of fifty-six feet. Without blow-out preventer equipment, we pulled out of the hole and lit the gas flare. A blow-out preventer is a mechanical device that attaches to the casing at the surface of the hole. If a dangerous occurrence arises, it allows the crew to shut off any back flow from the well.

We offset the hole location about one thousand feet and encountered a gas flow at seventy-five feet. After about ten days of futility we were forced to abandon the location. I was in the field for the ten days as we were awaiting an opportunity to log the well. Nola had a difficult time varying three meals a day, with sandwiches being her only option. The drill crew delivered my sandwich meals as they changed shifts. Fortunately, I had built a trailer complete with a space heater to house the logging machine. As the temperature set in at minus 55 degrees Fahrenheit, it afforded me a warm place to sleep and a place for the drill crew to warm up.

The cold weather and deep snow persisted but Nola and I agreed to travel to Moose Jaw for Christmas. As we had been married with very little notice, it would allow her mom an opportunity to introduce me to the relatives and friends in her home city.

We accepted a ride from Oyen to Rosetown, Saskatchewan, with fellow crew members, the McCartney brothers. They had been imbibing at the local bar and about ten miles from Oyen they rolled the car. Luckily, the snow was so deep that there was no damage to either the car or us and we were able to right it and carry on. I insisted on taking over the wheel and we arrived safely at our destination.

Our trip continued by bus to Swift Current where we were to catch a CPR passenger train to Moose Jaw. Because of the extreme cold and the snowfall the train did not arrive on time. The station agent continued to inform the waiting passengers of intended arrival times, which grew progressively later and later.

We finally rented a room at the hotel for a few hours. We departed at midnight, December 23, and arrived in Moose Jaw at 4 A.M., December 24. Mom Brown's live-in friend had been back and forth to the station half a dozen times.

The onerous trip was well worth the anguish as we had a great time visiting Nola's family and friends. Our return trip was rather uneventful. However, to schedule our ride to Oyen with the McCartney boys necessitated an overnight stay in the Rosetown hotel. As there was no running water in the hotel (very common in hotels in those days), our room was supplied with a ceramic water jug, washbasin, and towels. The washstand was next to the window, and in the morning the water in the jug was frozen.

One evening in early January we decided to do something different, so we dressed up and went to the bar. There were no seats available; however, a Big Rig crew was in town and one of the

Bob and Nola shortly after their marriage, November 1951. Bob proposed to Nola three days after they met.

drillers asked us over to their table. After the bar closed at 11 P.M. we were invited to continue our visit at the driller's trailer. In conversation, Cliff looked at Nola and said, "You're pregnant." Nola replied, " If I am, it has only been a few days. How can you tell?" "I can tell by the look in your eyes." As it turned out he was correct.

On January 8, 1952 (Nola's birthday), the crew received orders to move to Wilkie, Saskatchewan, a distance of 150 miles. The first night we stayed at the hotel and the next day hunted for affordable accommodation. A local suggested we contact Marion and Max Clement, who had just moved into a two-storey house and might wish to rent the upper floor. They were recently married and were delighted to have another young couple share their home. Funds for entertainment were sparse for both couples so cards and board games were the norm.

Their bathroom was complete with a washbasin, tub, and "chemical" toilet. A chemical toilet consisted of a five-gallon portable bucket housed in a metal stand with a toilet seat attached. Each time the bucket was emptied a small amount of chemical was poured in the bucket to hopefully mask the smell. When the bucket was nearly full you carried it through the house and dumped it into a holding tank in the alley. The "honey wagon" came about every two weeks and emptied the holding tank.

Nola had not confirmed her pregnancy but Marion was pregnant so the bucket filled more rapidly than normal. The rule was that the last person who used and filled the bucket was responsible to carry and dump the refuse into the holding tank. Usually the men performed this unsavoury task. A farm friend of the Clements stayed with us while his wife was in the hospital. Max and myself, working in unison, were able to stick him with the job of emptying the bucket about 90 percent of the time. We missed him after he left.

In February the local doctor confirmed Nola was pregnant. A

blizzard blew for two weeks and the drifts on some banked corners of highways were fourteen to fifteen feet high. Our stay in Wilkie was short lived and we moved the crew north of North Battleford to Glaslyn, Saskatchewan.

Our project covered the territory north of Glaslyn to Meadow Lake. The crew lived in the Glaslyn Hotel and ate our meals at the cafe or a boarding house. Glaslyn was a very small community. The hotel lacked central heating and a space heater at the end of the hall provided the heat for that floor. A diesel plant provided the town's power and ran from 6 A.M. to 10 P.M. It was pretty primitive living.

As usual, money was scarce, so Nola stayed with her cousin Phyl and husband Bruce. They lived in an apartment building that had been refurbished from a World War II RCAF bunkhouse. Both Nola (whose pregnancy was very evident) and Phyl were expecting, with Phyl's due date rapidly approaching. While Phyl was in the hospital having her baby, Nola provided meals and everyday chores for Bruce and their other child. Bruce travelled five days a week as a salesman.

I received word from the head office to go to Peace River, Alberta. A surveyor had resigned for no apparent reason. It was suspected that he was unable to tie a control line.

I visited Nola in North Battleford and flew to Peace River in a Ford tri-motor plane with burlap web seats (this was my first airplane flight). Gene Cook, who had been my party manager previously in Hondo and Camrose, met me. During the winter of 1951–52 there were fifty-eight seismic crews working or being supplied out of Peace River. The conditions in Peace River bordered on utter chaos. Sleeping accommodation was at a premium, and Gene, who was six feet, three inches, and myself at six feet, two inches, were forced to sleep together in a single bed. The room had an additional ten beds.

We arose at 2 A.M. and drove to camp, just west of Keg River Cabins (located on the Mackenzie Highway about 150 miles from Peace River). We arrived in camp about noon. This was a hot-shot crew consisting of a small complement of men including the cook and camp attendant. The camp moved nearly every day. The two bunk trailers and combined cook and wash trailer were mounted on Athey Wagons pulled by D2 or D4 Caterpillar tractors.

Fortunately, we had a fantastic cook and the best food available. Entertainment was at a premium so the crew decided an eating contest was in order. They picked another fellow and me as contestants. We ate seven large T-bone steaks and five soup bowls of ice cream smothered with frozen strawberries at one sitting. No winner was declared.

The Caterpillar tractors pulled the recording trailer and were used on the drilling crews. The drills were converted Boyles Brothers manufactured underground mine drills (auger) fitted to a D4 cat. My survey vehicle was a Fordson tractor with a bogey wheel and tracks. On a steep incline it had the propensity for rolling over backwards.

My first job was to find the suspected survey error. That did not take long. On a very long tie line through a valley the line declined and then rose to the other side. The negligent surveyor's readings continued to decline. Imperial Oil was about to commence with a drill sight until I found the error.

Communication had not improved from my previous experience in the north and our contact with IOL's Peace River office was sporadic. Our food and supplies were dropped from airplanes. We were completely isolated. The crew had been there since freeze-up in November and had very little contact with anyone other than camp personnel. Some of the crew members were feeling the effects of bush boredom. One in particular loved to hone and polish this big knife. One day he threw it across the bunk trailer narrowly missing another chap. We shipped him out.

This area of Alberta was completely blanketed with muskeg

and after the spring thaw started we had a devil of a time keeping the tractors from sinking to eternity. We did not get out of there until April 15.

In the spring of 1952, Nola and I rented a suite, as Northwest indicated that I would be working out of Calgary for an extended period of time. Apartments and suites were very difficult to find, particularly with a child in the offing. We rented a partially furnished basement suite near Kensington Road and Twenty-fourth Street (now Crowchild Trail). We purchased furniture and within a month I was transferred to Wilkie.

Bob with survey instruments, Medicine Hat, 1951. It was Medicine Hat where Bob would meet and marry his future wife, Nola Cooper.

We let the suite go, put what little furniture we had in storage, and took off for Wilkie in a company half-ton. We routed our trip through Moose Jaw so that a pregnant Nola could stay with her family. I continued on to Wilkie and was able to secure room and board in a boarding house operated by Nora Edwards.

Neither Nola nor I were happy living apart so I worked out a plan with Ron Elder to trade my room and board at Nora Edwards' for his home-made trailer, parked in Nora's backyard. It was twelve feet long and eight feet wide and consisted of a bed, a clothes closet, a half table (dropped from the wall), and a hot plate. There was no running water or a sewage system, which meant we had to use the refrigerator, stove, and bathroom in Nora's house. We hauled water for personal use and dishwashing.

I "borrowed" the survey pickup and brought Nola back to Wilkie. We "borrowed" some dishes and linens from Nola's mom. When preparing a meal we carried whatever food we required from Nora's kitchen, except cereal, bread, etc. If we were having a bowl of soup we used an electric hot plate, otherwise we used Nora's stove and oven. It was some feat for Nola to balance the full plates in her pregnant state (she was huge) out the back door of the house and up the step into the trailer.

However, we had a lot of good times in Wilkie, usually a party every weekend, and as Nora was a widow, she joined in the fun. It was house parties or a local dance. One of the drillers on the crew was from Ireland and apparently was one year away from finishing his medical degree. He had a great sense of humour but was rather unusual. The rig operated round the clock and he preferred the graveyard shift. This shift was not a favourite, so he traded easily with the other drillers. When his shift was completed at 8 A.M. he would come to town, wash up, eat breakfast, and head to the beer parlour for the 10 A.M. opening. He would quietly drink beer while reading a book until his evening meal. He would sleep until 11 P.M. and head out to work.

Tim loved his Irish whisky and on occasion would over-imbibe. One evening as we were preparing to leave for a dance, he was boasting about his prowess with the girls and passed out. No amount of shaking, yelling, or whatever would arouse him. He sported a very luxuriant mustache, so we decided to shave off one side of his mustache and the opposite eyebrow. Let's say he was not amused when he awoke the following morning.

While the rig was in Glaslyn the previous winter the power plant had given up the ghost, and we were lucky to be able to rent another from the local welder. We had repaired our original plant, though, and the rent on the borrowed one was adding up. One Saturday in June, Nola and I loaded it on the one ton and proceeded to Glaslyn, which is north of North Battleford. We had a flat, and I was in the process of changing the spare in some very warm weather when a fellow stopped. He did not offer to help but advised me that if we had a Saskatchewan Motor Club Membership we would not have to fix the flat. He then took off without offering help. He was very fortunate I was unable to catch him.

Nola and I visited the boarding house I frequented while working in Glaslyn. The lady who owned it wondered if we were coming up for the wedding. We were not aware of any wedding and asked who was getting married. She told us that Ford McCartney, one of our drillers, was marrying a local girl and there were guests coming from Europe. The wedding was scheduled within a couple of weeks and was the talk of the town.

We returned to Wilkie and on checking around found that no one seemed aware of a marriage. None of the crew questioned Ford on the matter and about three weeks later the father of the "bride" showed up with the unfortunate young woman. He confronted Ford and questioned him about his absence at the wedding. I have no idea what answer Ford offered, but the father advised Ford that the girl was now his responsibility and left her behind. The girl slept in Ford's car until we completed our contract and headed to Calgary.

Ford never offered an explanation to the crew but took the young woman along to Calgary. The Calgary Stampede was in full swing and the city was overrun with tourists. Subsequently, Nola and I stopped for a cup of coffee at the bus depot and here was the "bride" slinging hash. Ford had accompanied her to the Stampede and disappeared.

After arriving in Calgary I was asked to take over as survey supervisor. Our living accommodation was unsettled so we stayed at my parents' home. My first assignment entailed a survey school in Lethbridge, Alberta, to train a number of budding surveyors. One of the surveyor recruits was Dennis Draper, my second cousin.

Northwest had two seismic crews in Lethbridge, one managed

by Jack Williamson and the other by Don Carter. Don had been on my first crew at Bashaw and his sense of humour had not changed. We stayed at the Marquis, "the hotel" in Lethbridge. After work one day a few of my surveyor recruits and myself were having a beer in my room. Unbeknownst to me, my room was directly below Carter's room. He unrolled the fire hose, which was located in the hall, through his room and out the window. He then shinnied down the hose and into my room. The hotel management was not overly impressed with this little caper.

While Dennis and I were surveying a line south of Lethbridge, near a Hutterite colony, we stopped for a rest and a smoke. Shortly thereafter a couple of young Hutterite men showed up with a few bottles of homemade ice cold beer. They offered one bottle of beer for one cigarette. It was a very hot day and the offer was quickly accepted. Amazing as it may seem, we surveyed that same line for a number of days in a row.

After I had trained the required number of surveyors I was assigned to Foam Lake, Saskatchewan, along with Doug MacMillan to complete an E-log contract. Then I headed to Tompkins, Saskatchewan, to solve a survey problem with one of our crews.

Gold Rush Country

Then it was on to Barrhead, Alberta, to solve a survey problem on a crew contracted to Imperial Oil Limited. I conscripted Dennis as my rod man. The crew was working among the sand hills north of the Athabasca River. This was an interesting area as it was the site of the Canadian overland route to the Klondike Gold Rush in 1898. Edmonton was the supply point and the trail was long and difficult. It wound its way through the river systems down the Mackenzie-Liard Rivers and across the mountains to Dawson City. This route was very impractical and it was evident, due to the many gravesites along the trail, that many potential prospectors had not completed the trek.

This was my first experience with a Land Rover Jeep and I was sold from the beginning. This one had been built for World War II and could very nearly climb trees.

Norman (Chips) Cowper was the IOL survey supervisor who assisted me with our survey problem. We had been crew members with Northwest at High Level in 1950. His home base was Edmonton, which enabled him to spend weekends with his family. Nola was momentarily expecting our first child as was his wife expecting their first. I resented that he could to travel to Edmonton daily to see his wife while my company was not so lenient.

From Barrhead I returned to Calgary for the birth of our first child, Brenda Colleen, on August 31, 1952. Nola and I were visiting her uncle Ed and aunt Nora (he was the Indian Agent at Morley, about fifty miles west of Calgary). Nola was experiencing contractions so a Morley Hospital nurse suggested we return to Calgary. I dropped Nola at the Calgary General Hospital. In those days fathers were not allowed in the delivery room. As a result, I spent the rest of the evening visiting a friend.

Continuing to live with my parents was out of the question and although accommodation in Calgary at that time was difficult, particularly with a wee baby and a dog, we found a suite of rooms in the basement of the Waddell family home on Twenty-one A Street off Seventeenth Avenue SW. We had recently purchased a dog from my cousin in Taber. He was an English Cocker Spaniel named Kelly.

Northwest was awarded a seismic contract by NFA (Northern Foothills Agreement), a consortium of Shell, Gulf, and Texaco. On Boxing Day, 1952, the crew shipped out for Boundary Lake, northwest of Fort St. John, BC. When we arrived at Dawson Creek we were advised that the muskeg had not frozen sufficiently to support the dozers. We spent New Year's Eve, 1953, in Dawson and the crew had a merry time. We departed on January 2. The townspeople were not unhappy to see us leave.

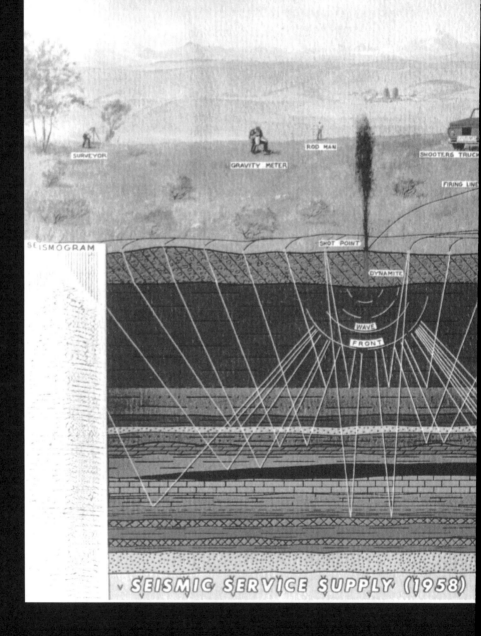

Seismic exploration, 1950s. *Valley Tool and Machine Ltd.*

SURVEYOR

GRAVITY METER

ROD MAN

SHOOTERS TRUCK

FIRING LINE

SEISMOGRAM

SHOT POINT

DYNAMITE

WAVE FRONT

SEISMIC SERVICE SUPPLY (1958)

That was "some crew." Except for the odd driller, I was the only one with bush experience. The rest of the crew consisted of young recruits from Lethbridge (at least I knew some of these young men from the survey school I ran the previous summer). The party manager was Harry Young, who was not versed in seismic field operations. He had joined one of our Lethbridge crews the previous summer and the party manager on that crew thought he would be a great candidate for party manager. I more or less ran the crew that winter as Harry was off daily to Fort St. John to ogle the CPA stewardesses.

It was an extremely cold winter with temperatures dipping as low as minus 72 degrees Fahrenheit. Our trucks were equipped with V8 motors complete with electric block heaters, one for each bank of four cylinders. The engine would start but the wheels would not rotate due to frozen bearing grease. We thawed the wheel bearing grease with a blow or propane torch.

Frozen juice was not available and the juice: tomato, orange, etc., was supplied in forty-eight-ounce tins. These young fellows drank up to three or four cans daily, as the alternative—although safe to drink—was muskeg water, which was a dirty brown colour. I warned them to temper their consumption of canned juice as it would bind them up tight as a drum. They did not take my advice and after a few days were doubled up in pain.

I was the surveyor on the crew and my rod man, Dennis Draper, was one of the prolific juice drinkers. We were running a control line one day and the survey line had a slight dip in the contour. I could see the top of the rod for survey shots. Dennis had the truck with him. All of a sudden the rod went down and there was no sign of Dennis. I threw the instrument tripod over my shoulder and walked the quarter mile and found him lying on the trail doubled up in agony. Lesson learned.

Although survey control was sketchy in the area, there were benchmarks located on the BC/Alberta border that gave us a starting point. I ran a control line up to the closest northern base line and attempted to tie to a Dominion Land Survey benchmark. I was dead on in vertical but out 750 feet in horizontal control. I resurveyed three times and came out with the same answer. Jack MacMillan, president of Northwest, was in camp with the client representative, Doran Holland. Jack said, "If you resurveyed the line three times, I'll take your survey over any Dominion Land Surveyor." His confidence in my survey abilities was very rewarding. It was suggested I contact the BC Land Titles and Survey Office in Pouce Coupe and look up the survey notes of the original surveyor. Sure enough he had recorded a 750-foot horizontal error at the base line in question without correction.

There was another crew working for NFA, Geophysical Associates of Canada (GAC), parked about five miles north of our camp. Their surveyor was very "green" and Don Burtt, the party manager, asked me to give him a hand locating a benchmark off the Alberta/BC border. I agreed and Don and I became close friends. Some time later over coffee, Don told me he was getting married in the spring. The crew had worked the previous fall in southeastern Alberta and domiciled in Medicine Hat. It is hard to believe, but Don was marrying Marion Byers, the girl whom I had taken out in Medicine Hat a couple of years previous and through whom I found Nola. The following spring they spent part of their honeymoon with us, and so sparked a close friendship that continued until their deaths: Marion in the early 1990s and Don in 2005.

The project was very remote and few if any humans had visited the area. The bush coyotes would eat out of your hand like a family pet. When eating lunch with the vehicle window open, whisky-jacks liked to perch on the steering wheel and mooch a morsel of food.

While Dennis and I were laying out a new line for the crew to shoot, the truck broke down. We were about six to seven miles from camp and Dennis walked to camp to get help. I waited about three hours and as he had not returned I decided to hike it to camp. There was a new skiff of snow from the previous evening and I was surprised to find a fresh set of large cougar tracks following Dennis's. On our return to the vehicle, Dennis was startled to see the tracks, as he had no indication he was being followed.

Our contract was completed in early April. After more than three months in the bush, the crew was pretty antsy to get home. The normal route was bush trail to Charlie Lake, gravel to Fort St. John, and Alaska Highway to Dawson Creek. From there we drove to Grande Prairie, Valleyview, High Prairie, Smith, and finally on paved roads from Westlock through Edmonton and on to Calgary. The Valleyview-Whitecourt cutoff was not completed until 1957.

We decided to shorten our trip by taking a bush trail from camp to the Alberta/BC border, then a gravel road to where the current Dunvegan Bridge is located. We crossed the ice on the Peace River at this point. Before attempting to cross we tested the ice and although the water was flowing on the surface of the ice, we decided it was safe. By "tested" I mean we drove the lightest truck across the river without incident, then the heavier ones, one at a time. We drove with the driver door open, and the sphincter muscles were pretty tight. We reduced our driving time to Calgary by half a day.

While at home with my family we entertained crew personnel, and as previously mentioned, Don and Marion Burtt on their honeymoon. During one of our hosted crew parties, I became fairly inebriated and went to bed about midnight, leaving Nola to entertain our guests. She was feeding our daughter every two hours and needed her rest. The last of the drunken group remained till 4 A.M. Nola has never let me forget one of my earlier married indiscretions.

Saskatchewan Times

After a few weeks working in the home office, finalizing reports, and maintaining and painting the equipment, we received word that we had a won a large contract with Mobil Oil in southern Saskatchewan. We left in late April 1953 for Assiniboia, Saskatchewan, approximately five hundred miles southeast of Calgary. Assiniboia was a very friendly town of approximately thirty-five hundred souls.

We retained our suite in Calgary because rental accommodation was at a premium and we were not certain when we might return. With our Calgary rental obligation, accommodation for a family in Assiniboia was out of the question. Nola, Brenda, and our dog stayed at her mother's home in Moose Jaw where we paid forty dollars per month for room and board. Nola's mother and stepdad worked and Nola was able to assist by making meals and housekeeping.

Dennis and I shared a room at fifty dollars per month. We ate at a Chinese restaurant. What with monthly suite rental in Calgary of seventy-five dollars, there was not a great deal of money left for my meals, clothes for the family, and a little entertainment. My gross salary was 225 dollars per month and with deductions it is easy to understand our circumstances. Al Campbell was married and in the same boat but he was making a little more money than I was.

Our breakfast consisted of two slices of toast, jam, and a glass of milk. The café made one sandwich for lunch. Dinner was a bowl of soup with crackers, butter, and ketchup. The crackers were supplied by the basket with a slab of butter. We concocted a meal called "prairie oysters," a cracker with a healthy portion of butter dished in the middle and filled with ketchup. Eventually the owner caught on and after about two weeks we received our soup, four crackers in a plastic wrap, and portioned amounts of butter patties.

We could not exist on the rationed portions so the circumstances encouraged Dennis and I to find a place to board. About half a block from our rooming house we found a family who would feed us three meals a day for forty dollars a month. This family had been on the farm during the "dirty thirties" and was thrifty to a fault. One day we opened our lunch bags to find tomato soup sandwiches: two pieces of bread, thinly buttered, with a thin layer of tomato soup. We suggested that she could afford some luncheon meat for our sandwiches and we had no more problems.

After a month or so it became apparent to me that we had a "drillers' union," where they only drilled an agreed-upon number of holes per day. We had three drill crews and were digging 150-foot holes. Each drill should have drilled five to six holes a day but the three rigs were producing a total of nine holes. The party manager, Harry Young from our previous contract at Boundary Lake, remained very naive. The drillers knew how to pull the wool over his eyes. I would spend two days a week surveying and five days in Moose Jaw with my family. I arranged rides back and forth with the local drill bit peddler, Leo Hurtibise.

In June, Harry returned from a head office visit and contacted me in Moose Jaw. He advised that he was moving to Calgary to work in the office and asked if I was interested in taking over as party manager with a healthy raise in pay to 275 dollars per month. It did not take me long to accept. I was twenty-three years old in charge of a seismic crew with a few hundred thousand dollars' worth of equipment and a crew of about twenty men. I continued my trips back and forth to Moose Jaw to see the family, but only on weekends. The promotion also meant I had a vehicle for my personal use and an expense account.

On returning to Assiniboia as party manager I had a discussion with the crew letting them know that I was aware of the "drillers' union" and I expected at least five holes per day per rig. I further advised that I would not stand for tardiness or lying. Either of these misdemeanours would result in the loss of their jobs.

Dennis took over as surveyor with Ken MacDonald as rod man. Both had been trainees of mine in Lethbridge. Len (Pinky) Morse was the operator; Jack Buchanan the junior observer; Daryl Richards (another Lethbridge trainee) was the reel truck driver; and there were about four others on the jug crew. The drill crew consisted of Mel Campbell and his brother Doug, with a 1000 Mayhew; Leo Grismer and his water jack with a 750 Failing (this was a contract rig owned jointly by Al Befus and George Blunden, at that time Northwest's chief interpreter); and another 1000 Mayhew whose driller and water jack I cannot recall but the driller was a French Canadian. The latter driller lied to me about winching the drill out of a ditch where he damaged the rig so it could not operate the next day. The rig was company-owned and I fired the driller and sent the rig back to Calgary. I hired a contract rig from Hussen Shibley in Radville and he sent Don Barrett as driller and Casper as the water jack. I told Barrett he had a job as long as he was the lead driller with the best daily production.

Our production went up dramatically and from that day on we were one of the highest production crews for Mobil Oil in the world! We remained under contract to Mobil Oil in southern Saskatchewan for three years. Another contractor in the area had three crews working for Mobil. They lost their contract and we picked up the extra work.

After we completed the required monthly hours as posted by the recording crew (220), the crew was able to take the rest of the month off as long as all the equipment was in good order to start the next month's work. I would finalize the monthly reports and ship them off to Regina where our computing office and Mobil Oil's regional office were located. Copies of monthly statistical reports were mailed to Mobil Oil's New York head office. The New York office requested computations to six decimal points. I had difficulty understanding the reason for this and contacted Mobil Oil's chief geophysicist in Regina. He agreed with me and instructed me to take computations to two decimal places in the future. It pays to question illogical demands when you feel the request is superfluous.

Usually, when our monthly hours were attained, I would drive with my family to Calgary for a head office meeting and return with supplies. This allowed us to visit with family and friends. Our route was from Assiniboia to Moose Jaw and Highway 1 to Calgary. All but a few miles on either side of Moose Jaw, Swift Current, Medicine Hat, and Calgary were gravel. As a matter of interest there still was not a road from the Suffield corner to Tilley. From Medicine Hat we routed through Lethbridge on Highway 3 to Fort Macleod and Highway 2 to Calgary. Our daughter, Brenda, was prone to car-sickness, which required us to badger some small town pharmacist into selling us anti-nausea pills without a prescription.

After the first summer we were extremely unhappy living apart and agreed to purchase a house trailer. I contacted the CIBC bank in Assiniboia to borrow the money for the purchase and was told there was no way they would loan money for a trailer because we would be moving all over. I suggested that was what a trailer was for ...

On our October 1953 month-end trip to Calgary, we purchased a used twenty-eight-foot-long (three feet of that was hitch) by eight-foot-wide Kit trailer for thirty-two hundred dollars. We had no money, our parents had no money, and the banks would not lend us any money, so I approached my uncle Gordon (my mother's youngest brother). I offered him 10 percent interest, which was higher than the going rate, and promised payment within three years. As it turned out, we scrimped and saved and repaid the loan in twenty-eight months.

We purchased the trailer from a couple living at the Bluebird

Motel and Trailer Park situated on the east side of Macleod Trail about one block south of the Stampeder Hotel.

Our trip from Calgary to Assiniboia was an adventure. By the time financial, licence, and other items were settled it was suppertime. I was unaware of the problems that manifest themselves when a parked unit has been stationary for some years. It never dawned on me that the clearance, stop, and brake lights needed maintenance. I did know enough to attach a hitch and an electric brake system for controlling the trailer brakes.

It was midnight of Halloween, October 31, 1953. We hooked up, loaded the family in the cab of the truck, and away we went. By now Highway 1 had been completed from Tilley to the Suffield corner, which allowed us a direct route to Moose Jaw. After some fine tuning of trailer problems we managed to stop in Brooks at about 3 A.M. After a couple of hours rest we continued and finally arrived in Assiniboia. Never too much work to have your family with you.

We were greenhorns in the operation of a house trailer. This particular unit was manufactured in California and the water and sewer system were exposed to the elements rather than insulated and enclosed in the sub-flooring. It did not take us long to wise up after the first big freeze when the water and sewer system froze solid. Using electric heater tape, manufactured specifically for such conditions, we became very cozy during the winter. We skirted the trailer from outer floor to ground with plywood and banked snow for insulation. The windows were single pane and we stretched and taped poly film to the inside, which formed a very satisfactory storm window. We used a hair dryer to firm up the poly film. We added a porch so the cold wind did not blow directly into the living area and it also provided much-needed storage.

We met a number of great townsfolk in Assiniboia. One family in particular was the Kyles. Norm was a partner and service manager at the Chrysler auto dealership. He was a first-rate mechanic and maintained our fleet of vehicles. His wife, Elaine, was a sweetie, and she and Nola got along famously. They had two children when we first met them, Gord and Bruce. Laurice, their daughter, was born about the same time as our Brent and she was named for my wife, Laurice Nola. Subsequently, a daughter, Joan, and a son, Kenny, were born. Along with other friends we attended picnics, curling, parties, and just plain visited. Some years later Bruce was in a horrific car accident that, ten years later, led to his death. We still visit the Kyles once a year. Unfortunately, Norm passed away in 2005.

There are so many stories to relate from this era, it is hard to know where to start. Our picnics were usually held on a Sunday; we chose a country location and our wives prepared fried chicken with potato salad, raw vegetables, pie, and watermelon. Elaine brought her fantastic homemade dinner rolls. We played ball and Indian wrestled. Indian wrestling consists of two fellows lying on their backs on the ground next to each other. At the count of three they would interlock their inner legs and attempt to flip each other. The winner continued competing against challengers until defeated. Those who usually attended the picnics were the Kyles and their offspring, Russ and Mary Carter, Laura and Bill Anderson, and Jimmy and Joyce Mitchell, among others.

We had some hairy experiences on the crew. We worked year-round on the prairies, which was unusual for seismic crews in those days. Normally, they operated in the far north during the winter. On one occasion, we were exploring an area in the badlands east of Assiniboia, which included the main east-west rail line through the area. One of our drills became mired in a deep snowdrift and the only available winch location was the railway tracks. The drill and truck were extremely heavy and I was not apprised of the indiscretion until after the arrival of the next train into Assiniboia. It arrived without incident.

During the winter of 1953–54 we transferred to a prospect in the badlands southwest of Rockglen, near the U.S. border (Rock-glen is about forty miles south of Assiniboia). Our dozers opened the roads allowing the local ranchers and farmers to replenish their larders. Our operator was ill and I was acting in his capacity. One afternoon, huge snow-laden clouds were observed rapidly closing in on us, the foreboding of a very severe blizzard. I dispatched one of the jug hustlers to warn the drills to complete the holes they were on and head for home. We shot whatever holes we could and headed for home ourselves.

The blizzard hit and engulfed us with blinding snow and severe winds. It was impossible to reckon the road surface limits, requiring a couple of the crew to proceed by foot in advance of the recording truck. This helped eliminate the possibility of sliding into the ditch.

The road was rapidly drifting in and it was obvious we would be unable to reach the highway, some five miles away. Around 5 P.M. we arrived at a farmhouse where they welcomed the ten of us, who constituted the recording crew and dozer operator. The blizzard blew for three days and fortunately, due to the roads being open, the family's larders had been replenished. Otherwise, with the additional ten people, rations would have been extremely skimpy. Near midnight we heard a faint knock on the door and opened it to find one of the driller's helpers. He had been unable to reach the highway as a three-foot drift had engulfed his three-ton water truck. He had followed the fence line for five miles to the farm.

To while away the time the boys played poker and the farmer asked to be included in the game. He lost a fair amount of money and I included his losses with the funds compensating them for food and lodging. Finally, the blizzard broke and we followed the dozer to the highway. When we arrived at the drill helper's truck, the cab was full of hard-packed snow. We had to break the snow into workable chunks with shovels for removal. Apparently, this particular truck had been located in Peace River the previous winter and had been repaired after a roll over. Peace River body shops left a lot to be desired and the driver's side window could not be closed tightly. The size of the opening was minute yet the snow had packed into the cab. That gives you some idea of the ferocity of the blizzard.

Apparently, on learning of our predicament, Mobil Oil sent out an airplane to attempt to locate us. But because of the continuing blizzard, they had abandoned the hunt. They were reassured by the drill crew members who had made it to town that we would be holed up at the farm.

Following this escapade, I felt the crew needed a break, so we drove across the U.S. border to Glasgow, Montana, for a couple of days of R & R. We certainly had a good time and escaped town before anyone ended up in the hoose gow (jail).

The following winter we were working east of Willow Bunch when another blizzard hit. I was with the crew and we were able to shut down in time to find accommodation. Six of us stayed at a farmer's house for one night. The prairies are not the place to carry out geophysical work in the winter.

We operated in south central Saskatchewan for three years. Assiniboia was the hub and we domiciled in Rockglen, Willowbunch, Moose Jaw, and Bromhead for shorter periods of time. We spent the better part of a fall in Willow Bunch, where we were able to park our wee trailer in a vacant lot next to the mayor's house. As there were no water or sewage facilities for our trailer, they allowed us the use of their washing and toilet facilities.

Willow Bunch was the original location of a large Métis colony (established in 1870) and very few adults spoke English. French was the language so Nola shopped after school, when the teenagers who were bilingual had part-time jobs in the grocery store. Nola, while washing clothes, would attempt to talk to the mayor's budgie bird. After a couple of weeks it dawned on her that the bird spoke only French.

Our son was born on December 12, 1954, when we were living in Moose Jaw. When Nola was near her delivery date, she travelled to Calgary and stayed with my parents. A phone call advising me of Brent's birth arrived around 5 P.M. I drove all night and arrived in Calgary at 7 A.M. to greet Nola and our new bouncing baby boy, Robert Brent.

We remained through Christmas and New Year's and arrived at our tiny trailer in early January 1955. It was so exciting to have our family together. I had worked hard to clean and tidy the trailer but when we opened the door a flood of water greeted us. A pipe had frozen and burst and the water froze to the floor. Fortunately, the trailer was set up with a slight incline to the rear where our bedroom was located and the excess water escaped through a crack between the inner floor and the wall. We left the children with our neighbours and proceeded to mop up the mess. Luckily, there was very little damage as the floor covering was linoleum.

The following spring after the thaw we were on the move to Bromhead. When I removed the skirting from the trailer there was a huge hole underneath. I could not possibly tow the trailer without it falling into the hole. We hitched our neighbour's Jeep to the trailer and with careful maneuvering we made a clean getaway without losing the trailer.

The population of Bromhead was twenty-five souls. There was a grocery store and a hotel run by Johnny and Clare Franks. John liked his booze and Clare ran the place. She was a great cook and as this was a hot shot job (hot shot is where the crew have all their living expenses paid), the crew lived and ate at the hotel. We parked our trailer adjacent to a bar window and threaded our power cord to an inside electrical outlet. Daily, when the crew arrived back in town, I joined the boys for a beer. When dinner was ready Nola would call through the window.

Mobil Oil had a "farm out" from Central Leduc. CL wanted

Bob, Nola, and children, Brenda and Brent, with Northwest Seismic's station wagon at Twin Bridges on the outskirts of Calgary, 1955.

and Mobil was hesitant to drill an exploration hole and we were to complete our survey as rapidly as possible.

This area in Saskatchewan was not under 100 percent road ban, but eventually the hammer dropped and we were shut down. I was anxious to get the crew back to work as the men were getting out of hand with little entertainment in town except the bar. Johnny Franks knew the reeve and council members of the municipality and all we had to do was load up a couple of bottles of rye and a few cases of beer and visit them individually.

This system worked like a charm and by midnight we had the okay to go back to work. I wanted the crew to commence work in the morning but had to round up one of our contract drill crews. We continued on to Radville to notify Hussen Shibley, who owned one of our contract drills. He was hosting a house party.

A few more drinks and we carried on to Milestone with the driller to pick up his helper. The helper was living in a small trailer with about six kids. They were sleeping on the floor crosswise like cord wood. On the way to Milestone Johnny Franks was unfit to drive due to overindulgence and I thought I was okay. Obviously I was wrong, as I was in and out of the ditch about five times.

We returned to Bromhead, and found that all passengers were accounted for and in one piece. It was now 7 A.M. and we saw the crew off to work. I stumbled into our trailer with the comment, "I'm going to Regina to pick up the client representative." Nola, the intelligent one in our family, said, "You had better eat," and provided a whopping breakfast of eggs and bacon. She knew, considering my condition, that after I ate I would pass out. Subsequently, that afternoon, I travelled to Regina and returned with the client representative.

We returned to Assiniboia for a short while until the contract with Mobil was complete. The crew returned to Calgary and I assisted in finalizing the reports for our three-year contract with Mobil Oil. I worked with our crew interpreter, and it was his responsibility to deliver the completed reports to Mobil Oil. Later that summer the chief geophysicist for Mobil, Gordon Gibson, contacted me inquiring about the reports. I advised him that Jim had delivered them. After an extensive search by Mobil Oil, the reports could not be located. We started our own investigation. I found Jim in a seedy Calgary hotel completely "out of it" in an alcoholic stupor and he insisted he had taken the reports to Mobil Oil.

We knew we were in trouble and sent one of our people to Regina Beach where Jim's permanent home was located. His wife had not seen Jim for a couple of weeks and said she thought there were some papers in the bottom of an old musty trunk, and on investigation they found the reports.

Opportunity Knocks

Due to the scarcity of trailer court spaces in Calgary, we parked our trailer at my grandpa Burke's farm, which was now owned by my uncle Mid. We were able to locate a spot near Motel Village across from McMahon Stadium (not built at that time, 1955). We stayed there for a short while and then moved to the Chinook Trailer Park on Macleod Trail just past Fifty-eighth Avenue. At that time the Chinook Drive In and Skyline Dine and Dance (currently the Chinook Mall) were immediately across Macleod Trail.

The company did not provide a vehicle but there was a bus loop at Fifty-sixth Avenue. We were fortunate to be adjacent to super trailer mates, Bob and Cathy Wilson and Nora and Ray Baynon. The three couples were not flush with money. A Saturday night party consisted of homemade fudge and a bottle or two of the cheapest wine available, which in those days was Mogen David. One weekend we decided to purchase a couple of bottles of wine. Wilson's dad had given him a car and with a near-empty fuel tank we proceeded to the nearest Alberta Liquor Store at Seventeenth Avenue and Eleventh Street SW. When we arrived at the liquor store we realized all of us had forgotten our liquor permits (in those days an Alberta liquor permit was required to purchase booze).

With not enough gas to return home and get one of our licences, what to do? "Well, I was raised in Calgary," I said. "Surely somebody will show up that will recognize me." Sure enough along came an old school acquaintance. I told him our sad story and here is three dollars, would he buy us a couple of bottles of Mogen David. He agreed and a few minutes later handed over the two bottles. I often wondered what his reaction was to my request (Mogen David was the wine of choice at that time for the local winos). Probably told our mutual acquaintances about how "old Rintoul" had turned into a wino.

The oil industry was in a slump and geophysical contracts

were hard to come by. Northwest dropped from fifteen crews to five. Part of my summer was spent in the Turner Valley oil field, E-logging water wells for the first water-flood operation (to the best of my knowledge) in Canada. Anderson Drilling dug the water wells and I logged them to ascertain the most prolific water-bearing sub strata. The procedure was simple: pump the water to a producing well and force the remaining oil to the surface. When I returned to Calgary after a day's work, I would consult with George Blundun, our chief geophysicist. (He eventually became exploration manager for Home Oil.) We would map the sub surface water-bearing strata and pass on the information so the oil company could commence water flooding as soon as possible.

After the completion of that contract, I took a crew, which among others was primarily made up of a number of out-of-work party managers, to Stettler. When we arrived, the talk of the town was about a fellow who, while attempting a left-hand turn, had his arm ripped off by an on-coming car. In those days, few vehicles were outfitted with turn signal lights so the driver extended his arm full length for a left turn and a "crooked" arm, with the elbow bent upwards, for a right turn.

During this oil industry downturn, I also helped on a number of "well shoots." This was a system of providing a check against geological information on holes drilled for oil or gas. When the exploration hole was nearing bottom, the oil company contacted our office and we immediately drove to the well site. The crew consisted of a driller and helper, surveyor, and a recording truck operator and helper.

Invariably, when we arrived at the rig, they had run into a problem and we had to wait for completion of the hole. The surveyor established the location of the drill holes, usually five or six in a semi-circle equidistant from the bore hole. These holes were drilled to about sixty feet and individually loaded with an explosives charge. A special geophone was lowered in the bore hole and at each horizon (changes in sub strata rock as determined by the core samples), the charge was detonated. A recording truck recorded the results of the energy and a log was produced. The log usually confirmed the geologist's recording of the different horizons.

A couple of unusual incidences happened while on, or travelling to or from, the well sites. One occurred when we were returning from a well shoot near Wabasca, east of Peace River. It was an extremely cold winter morning, about 4 A.M. My vehicle was dangerously low on fuel and I used high-test gas from our lanterns to supplement what little gas we had in the tank. On the way down the steep incline into Peace River, with the truck in the lowest gear, the motor was "pinging" so loud it sounded like somebody was shooting a machine gun. In the end, there was no damage to the motor of the truck. But even so, better a damaged motor than a couple of frozen bodies.

On another occasion, I had been on the road for over twenty-four hours and arrived at the rig site before the rest of the crew. It was a beautiful summer day and I laid out flat under the one ton to snatch some much needed sleep. During my sleep, a thunderstorm blew through the area and, as the truck was parked on a slight slope, the water woke me up. It was running in the top of my coveralls and out the legs. I could really sleep soundly in my younger days. During the winter, while waiting for the rig to bottom out, we often slept on a two-by-ten plank in the boiler room. *That* was noisy.

Nola and I traded our Kit trailer for a new Schult. It was forty-two feet long and eight feet wide. This unit had all the latest equipment: living room in the front; kitchen amidships; two bunk beds adjacent to the hallway; bathroom with vanity mirror, table and drawers, bathtub, shower, toilet, and sink; and the master bedroom in the rear. We were in seventh heaven.

During this break in fieldwork, B.J. Seaman of Sedco Drilling approached me to head up their slim-hole operation.

A house trailer was home to Bob and Nola and their children for years as they moved from place to place within the oil patch. Living in a trailer meant that they did not have to be separated. This was their second trailer, 1954.

Men with slim-hole drilling experience were at a premium. B.J.'s offer was very attractive and I told him I would think about it and contact him.

Later that day I visited Northwest's office and the president, Jack MacMillan, offered me a job on a new slim-hole contract in Jamaica as surveyor and E-log operator. Apparently, Stanolind Oil, which eventually became Amoco, was sending a Big Rig, a slim hole, and a geophysical crew to Jamaica for six months with a budget of 3 million dollars (about 23 million in today's currency).

Employee expenses would be paid and salary deposited in Canada. Families would not be accommodated for the six-month period. However, if we proved up a potential oil or gas find the contract would be extended for an additional three years. In the event of an extended contract, families would join us with all expenses paid. After a lengthy discussion, Nola and I and agreed to accept the proposition. It would allow us the opportunity to develop a "nest egg." I did not leave for Jamaica until February 1956.

In the fall of 1955 we took a crew to Bashaw (where I first started in the seismic business). We left on a beautiful day, complete with family and new trailer. We were able to travel over the new four-lane on Highway 2 starting north of Crossfield (still gravel) and on to Ponoka where we headed east to Bashaw. We arrived near dark and set up the trailer beside the house of Don and Gladys Whitney. Don and his father owned the garage where we purchased our fuel and they agreed to look after our mechanical problems. A blizzard arrived during the night with such severity that it blew the trailer fuel tank off the stand. I had not blown the water lines down and our toilet bowl water froze solid.

By morning the storm had abated, allowing the crew to start the new project. I did not get around to thawing out the trailer until later in the day. The ice had cracked the bowl of the toilet. Money was scarce and we could not afford a new toilet so we removed the bowl. We glued the pieces together and painted it with fast drying white paint. It was necessary to repaint about every six months but that toilet was in the trailer when we sold it some years later.

Our contract was with Home Oil and our crew was very experienced. We worked cross-country and all shooting lines had to be permitted from farm owners and the snow removed by bulldozers for easy travel. Northwest was short of help that winter. As well as my regular duties managing the crew, I was required to carry out permitting and surveying duties. Our contract provided payment for the three positions. When I heard that a Home Oil supervisor was about to visit I made certain I was either in the office or visiting the recording or drill crews. One supervisor arrived unannounced and the jig was up. We were required to place a permit man and surveyor on the crew.

This was the start of "pattern shooting." This entailed drilling numerous holes, up to twenty-five or thirty per shot point, loaded with very small explosives charges. As this system was in its infancy, it required a lot of experimentation, so you never knew from day to day how many holes per shot point were required. I permitted for the usual shot point of one hole, as that was what the farmers understood. However, when I left the crew for Jamaica in February 1956, the new party manager had some explaining to do to the farmers regarding all these holes per shot point.

We spent Christmas and New Year's in Bashaw and the "locals" treated us very well, lots of parties and a warm festive season. One of our fellows had recently returned from Trinidad with a new wife and she nearly froze to death. It was her first experience with snow and cold weather.

Our daughter, Brenda, was three-and-a-half years old and liked visiting. She would leave the trailer, knock on a door, and when it was answered would explain, "I am Brenda Rintoul. I live in that yellow trailer at the Whitneys, and I would like to visit." She always received a welcome. One day she came to me and said, "When are we moving?" I asked, "Why do you want to know when we are leaving?" and she replied, "Because I know everybody in town." It was necessary for children of "doodlebuggers" to be outgoing, as the rapid moves did not allow for lasting relationships.

To this point, my experience and life lessons had been local in nature. The move to Jamaica was tremendously exciting and I looked forward to expanding my horizons. I had no idea that it would so profoundly change my outlook on life. ■

Part Three

Life in the Larger World

Jamaica Bound

I finally received word in February to make plans to head to Jamaica. I had insisted on a two-week holiday in Ottawa to visit my parents. We pulled the trailer to Moose Jaw and parked it in Nola's parents' backyard. This is where Nola and the children would live until I rejoined them or took them to Jamaica. I drove the pickup to Calgary and immediately flew to Ottawa where I spent a pleasant two weeks visiting my parents.

I boarded a plane in Ottawa and flew to Toronto. Here I changed to a North Star plane and off I went to Jamaica. North Stars were a two-prop plane and very noisy. It was necessary to land in the Grand Bahamas to refuel after leaving Ottawa at 7 A.M. We arrived at the Kingston, Jamaica, airport about 9 P.M. In Jamaica, year-round, daylight arrives at 7 A.M. as if a curtain were raised and nighttime comes at 7 P.M. as if a curtain were lowered—no dawn or twilight.

Jack MacMillan was in Jamaica but sent a driver to pick me up at the airport. When I arrived at customs they asked me why I was visiting Jamaica and I told them I was here to work on oil exploration. They asked for my work visa. I had no such document. Then they asked me if I had a return airline ticket. At that point the driver came over, explained the situation, and I was allowed to enter the country.

What a physical and mental shock, after leaving the winter of the prairies and eastern Canada, to experience the warmth and humidity of a tropical country. I was twenty-six years old and had never been outside western Canada. Before I joined the geophysical industry I had never even been beyond Banff, Lethbridge, or Edmonton.

Jack had instructed the driver to show me around Kingston (Jamaica's capital) and then deliver me to the Myrtle Bank Hotel. (We subsequently nicknamed it the "Turtle Tank.") This was "the hotel" in Kingston. I had a beautiful room. I changed my clothes and joined a throng of people in the garden, which had tables loaded with lobster, shrimp, and crab. I ordered a Jamaica punch drink and charged it to my room. I had some Jamaica notes and change in my pocket and the drink was equivalent to about one dollar Canadian. I gave the waiter a nice shiny silver coin about the size of a dime as a tip. The waiter said, "A thrupence, sir?" I immediately dug deeper for some more coin.

I slept through the following day and awakened on the second morning. On arising I had breakfast and sauntered over to the pool where Jack MacMillan, Stew King, Nola Nickle, and a couple of other people were sunning themselves. It was interesting meeting Nola Nickle; she was the first person I had ever encountered with the same given name as my wife.

Before Jack and I headed to the west end of the island we visited Jamaica Transport, the official Jamaican rum truckers. The company also managed the aged barrels of rum for customs. The owner was Tinker Reary and his right-hand man was Joe Kiefer. They were the trucking firm handling our supplies to drill locations.

I required a Jamaican driver's licence, and Joe took me to the government offices to take a driver's test. Before the test began, Joe slipped the examiner a twenty-pound note. I was driving, the examiner was in the front passenger seat, and Joe sat in the back. Eventually we came to a "roundabout." I had never experienced one of these in Canada and the instructor asked, "Who has the right away in a roundabout?" I glibly answered, "I guess the first one who gets there." He looked at me rather oddly, then said, "You pass." I subsequently found out that in Jamaica, if there were a long straight stretch of road (which is rare) they would put in a roundabout to slow you down. They also had roundabouts at all intersections. Years later Calgary and Edmonton built a few roundabouts, and while in Scotland we experienced many roundabouts. But the Jamaican one was my first.

As Jamaica Transport was the customs custodian for the rum, it had empty kegs that were returned after the rum had been bottled. The dregs in the keg were very powerful, close to 100 percent over proof. It was a great trick to get a greenhorn like me to sample this libation. It was very strong and one drink was enough to affect your motor skills. Their rums were all of high alcohol content, particularly the white rum. Rum exported to Canada was usually cut 50/50.

The next morning Jack and I travelled to Little London just beyond Savannah-la-Mar, also called Sav-la-Mar, about 135 miles west and south of Kingston. Due to the mountainous nature of the country and the winding roads it took us about five to six hours to arrive at our destination. Jack was ecstatic with the crew's accommodation, but it did not impress me at all. It was an unpainted small abode with a raised cistern that supplied water by gravity. The water was unheated and the shower was inadequate.

The toilet flushed by means of a chain attached to the toilet tank, which was near the ceiling. Pull the chain and down came the water through a pipe to the toilet bowl. Where the bowl's remains went after that, I had no idea.

The rest of the crew consisted of three drillers and a toolpush. Luther Lakevold, the toolpush, and his family had been there for a couple of weeks. Luther, Jean (his wife), and their two boys lived in a neat little house near Savannah-la-Mar. Luther had worked with one of our slim-hole crews in Trinidad for a couple of years and had been transferred to Jamaica. We had a number of servants, including a cook, maid, yard boy, and mechanic. They lived in another building on the property. Prior to our renting the house, it had been lived in by a very wealthy Jamaican while his family home was being renovated.

About ten miles west was Negril Beach, owned by the West Indian Sugar Company. They kindly offered us the use of a portion of the most fabulous beach in existence, free of charge. There was an elderly coloured fellow living nearby and we gave him a weekly bottle of rum to keep the beach clear of whatever had drifted in from the ocean. Some time later I would swim from the beach over a reef one-half mile off shore and join one of my Jamaican friends for a pleasant Sunday sail.

Jamaica was a wonderful place, especially to a twenty-six year old who had not travelled extensively. Before leaving Canada I had visited the local library and learned everything I could about Jamaica, the topography, and its people. That helped me to understand the Jamaican people who lived in abject poverty. I became friends with many Jamaicans and they treated me as an equal.

Our drill crews consisted of a white driller and three coloured helpers. The helpers were paid the equivalent of thirty-five cents Canadian per hour, which was a very good wage in Jamaica. The rig worked around the clock in three shifts and shut down at noon on Saturday, then cranked back up at midnight, Sunday. That gave the Canadian crew an opportunity to explore Jamaica. Montego Bay was thirty-five miles north and was a favourite weekend get-away where we were able to procure the most succulent curried goat imaginable. Before we left Savannah-la-Mar for Montego Bay we would sit around George Lynn's store and drink white rum chased by Red Stripe beer and eat Jamaican meat pies. That was living at its best!

My outlook on life changed dramatically while in Jamaica. I had never been around other ethnic groups and their approach to life was so different from mine. They taught me a lot about life and I returned to Canada an infinitely more complete person than when I left.

In March, I commenced looking for a better place to quarter the crew and found a beautiful beach home complete with four bedrooms, living room, pantry, outdoor kitchen, and a tennis court. The house was located about fifty feet from the ocean on a protected bay and the local golf course was within walking distance. What a heavenly spot, and it was also near Luther and Jean's home.

I renewed my love affair with golf and was fortunate enough to purchase a set of Spalding clubs owned by a colonel in the British Army who had recently passed away. The Savannah-la-Mar Golf and Country Club, adjacent to our new home, was a nine hole course with a lovely little clubhouse consisting of a bar and billiard room. The barkeep was a fellow named Pilner and he looked after our every need. On Sundays, I would meet Osmond Hudson (the local Westmoreland Parish secretary) around 9 A.M. and we would play nine holes of golf. The weather was normally very hot and humid. After two holes of play you would be soaked with perspiration and sweat would be sloshing in your shoes.

The golf course was built on farm land and the clubhouse was built adjacent to the original farmhouse, cattle barns, and corrals.

The property was cross-fenced with three-foot stone walls, and the cattle grazed the fairways. The greens were fenced off to keep the cattle at bay. The first two holes ranged away from the clubhouse and the next two reversed, where Pilner would deliver a couple of very cold Red Stripe beers on a silver platter. He repeated the service at the end of seven holes and at the conclusion of our game. The course routed you over corrals and stone fences. It was a lot of fun to play.

There were two rainy seasons on the island, one in the fall and the other in the spring. I missed the fall rain but was certainly there for the spring rain. The rainy season lasted for about two months, and every day at precisely 4 P.M. it would, in one hour, dump ten inches of water. It would then clear up and be sunny before dark. The rain did not stop the golfers. The bridges had a vertical rod on the entrance, marked off in feet and inches so you could gauge whether it was safe to cross. Water flowed over the bridge in great abundance. The island was underlain with cavernous limestone so the water escaped rapidly. An hour later you would never know it had rained.

Life appeared very simple for the locals and when the mangos were in season it was difficult to convince them to work. Mangos, if eaten in abundance, have a lethargic affect on the consumers, leading to long sleeps in the middle of the day. We were able to keep our rig helpers away from the mangos because of the excellent pay they were receiving. Louise, our cook, and her sister, the maid, were not familiar with Canadian food so I bought them a cookbook. That was great except we ate in abundance the food described in whatever chapter Louise was reading.

One day, when Fred (one of the drillers) and I were sunning ourselves I turned to him and accused him of not smelling too sweet. He returned the compliment. We both smelled of a sickly sweet odour. We investigated and found out that Louise was cooking with coconut oil (a common practice among the locals). We immediately changed her cooking oil to butter. During crab season, the crabs would pull themselves out of the ocean and crawl under our beach home. We would trap, cage, and feed them our table scraps. In a couple of weeks Louise would clean them and we would have the most wonderful crab on the half-shell you could imagine.

In April, for a short period of time, we moved the rig farther east and stayed at a vacation motel on the beach: the accommodations were very lovely. The motel had a fresh water pool as well as the ocean a few steps away. We were in residence over Easter. The locals and some wealthy Kingston folks had a great time. They organized donkey races on the beach complete with pari-mutual betting. They talked me into jockeying one of these animals. The man I had as my servant taught me how to ride. The backbone of a donkey is razor sharp so you sat as far back on his rump as possible. A donkey command consisted of doubling his tail in one hand and squeezing when you wanted him to speed up. My feet were nearly dragging on the sand but I won three races, and as I had bet on myself, I won a substantial amount of money.

There were only ten white families in the area near us and we became great friends. They treated us royally. One family, the Aguilars, was most gracious and invited us weekly to their home for dinner. They had a grown family of four offspring: a single daughter, two married daughters, and a son who unfortunately was useless. One day Winston, the father, asked me if I would be interested in taking over his business without investment on my part. He owned a Ford motor car agency in Sav-la-Mar. He was prepared to sign the company and working capital over to me if I would guarantee to look after his family once he died (he was not a well man, with advanced malaria). It was very tempting but I balked at the ramifications.

During one of our conversations he asked me if I knew a lady from Canada who had married a Jamaican fellow. Canada is a big country but when he suggested she may have come from

Calgary and told me her maiden name, I immediately recognized it as the sister of the Calgary *Herald*'s ski editor (an acquaintance of mine). In 1956 travel was not as extensive as now, but it sure is a small world.

I was responsible for keeping the larder full. Our only source of meat was at the local market, which killed on Friday and served on Saturday, blow flies and all. I made twice-monthly trips to Kingston to purchase more edible groceries but at a premium price. Once I had loaded the meat and groceries it was necessary to skedaddle home before the meat thawed. The trip was 132 miles one way and took six hours to navigate because of the winding mountain roads. Front tires were replaced every five thousand miles.

Bob in Jamaica at Treasure Beach Hotel, 1956. His sojourn to Jamaica would change his life.

Along the route to Kingston, roadside vendors offered wonderful oranges and freshly roasted cashew nuts. I would eat these for lunch, along with a can of sardines and water biscuits. I stayed in Kingston overnight, loaded the vehicle the next morning, and headed home. One time on the way home, I did not leave Kingston until 5 P.M. I arrived at Thigpen, a small town on the route, in darkness. The government elections were underway and the animosity between the two political factions was fierce. There were many shootings and plenty of general turmoil. It was necessary for me to take the main street through town to continue west to our home. When I arrived at the edge of the village a policeman rode in the vehicle with a drawn pistol. On either side of the main road were opposing political factions straining to get at each other through a police cordon. As the policeman said, "Imagine what would happen to a white man if they broke the cordon and you were in the middle?"

Hill gangs barricaded the roads and when you got out of your vehicle to clear the way, they would let you have it with sawed-off shotguns. One evening about 2:30 A.M., Luther and I were returning to Sav-la-Mar after visiting the rig and came upon a barricade. Luther ran the barricade and we arrived home safely with two bent wheel rims.

I required a helper to act as my rod man, clear bush, and be a general all-round "gopher." I put out the word in the community and had fifty or more applicants. I picked a chap by the name of James who was forty-eight years old, legally married, and had a grade-three education, a rarity among the populace. He was a genuinely delightful person and it seemed he would do anything for "Mr. Bob." One day while picking up groceries in Sav-la-Mar, I offered James an ice cream cone. I asked him if he had ever had an ice cream cone and he replied, "Oh yes, Mr. Bob." I brought him his ice cream while he was sitting in the truck and I returned to complete my purchases. When I got back, here was James with his head and right arm out the vehicle window and the ice cream running down his arm and off the elbow onto the street. "James, you've never had an ice cream cone." He hung his head and admitted, "No, Mr. Bob." I showed him how to lick the ice cream cone in a circular motion to avoid the drips and he was most thankful. Can you imagine, a person who has never had an ice cream cone and is forty-eight years old?

During World War II the British Army surveyed Jamaica, establishing elevation and horizontal benchmarks, primarily on the peaks of mountains. Surveying was relatively easy in Jamaica as you could pick up zero elevation on the ocean and tie into one of these benchmarks. As Jamaica is tropical in nature the growth quickly overran the mountain top benchmarks. One day I carried an elevation as far up a mountain as I cared to go and sent James to the top to hack out the bush with the ever-present machete and give me a reading with the survey rod. It was going to take him a fair amount of time so I lay down and had a nap. After some time I had the impression I was being watched and looked up to find five vultures circling over me. I did not move and kept one eye partially open. They circled lower and lower and when they were about two feet above me I jumped up and off they flew.

You soon learned that as a white man you were a minority in a 99 percent coloured country. On one occasion I was travelling down a very steep mountainous road and met a huge truck coming up the grade. I pulled over to let him pass and he ran into me. Within five minutes a crowd of locals appeared out of the jungle, with everyone talking at once. A policeman was sent for and within a half an hour a very black man, partly dressed in a police uniform, arrived on his bicycle. We had disturbed his afternoon rest. He looked the situation over, talked to a few locals, asked me if I had blown my horn (a must in Jamaica if you expect any hope of being in "the right"), and then declared that no one was at fault. Just then our day shift arrived on their way to town. Our white driller was a very loud and sometimes obnoxious fellow and he started to berate the policeman about his decision. I pulled the driller aside and politely suggested he back off. I had visions of me residing in a very unpleasant Jamaican police cell.

When banking, the black clerks used any excuse available to serve a fellow black person before a white.

One of our drill helpers severed a finger and I took him to a doctor in Sav-la-Mar. I was appalled at the state of the doctor's office. The doctor came out of his office after examining a young pregnant Jamaican woman and threw the forceps in a cardboard box. There were oodles of flies hanging around.

I observed that Jamaican people are very superstitious and believe in the occult. They have what they call duppies (ghosts) and believe strongly in them. When somebody died, they held a three-day wake and chanted, moaned, drank rum, smoked ganga (marijuana), and generally threw a party. The participants of a wake may be heard many miles from the location. It was a rather eerie experience from my perspective.

On another occasion, Louise, our cook, had taken one of my shirts (a plaid one that Nola had made for me) to her quarters to sew a button on the sleeve. Her boyfriend saw the shirt, found out it was mine, and accused her of having an affair with me. He took his machete and cut my shirt into ribbons. If you could visualize Louise, a skinny woman with no teeth (her nickname, among us, was "gummy"), you would realize that there was no way Louise and I could have had an affair.

They had different ways of meting out justice in Jamaica. A fellow who killed another fellow over a girl would receive a six-month jail sentence but a fellow who stole a loaf of bread would get five years. I asked a judge about this and he explained that the fellow who killed another fellow over a girl was in a fit of passion but the bread stealer perpetuated a premeditated crime. It was difficult for me to understand but that was the way it was in Jamaica at that time.

On returning from the resort motel to our permanent house for supplies, I had an interesting experience. After the staff had served my supper, I read a little and jumped into bed. The bed was crawling. My immediate reaction was to think that the staff had been sleeping in our quarters (against strict orders), as I assumed the bed was full of bed bugs. I got out of bed, turned the light on, peered under the sheets and nothing. I crawled back into bed, turned the light off and bingo, more crawling. I got up a third time, procured a flashlight, crawled into bed, pulled the covers over my head and turned on the flashlight and for a fleeting second saw something scatter under the mattress. I looked under the mattress and the area was infested with ants. I sprayed the area and went back to bed for a peaceful sleep. I subsequently learned that it was flying ant season. They migrated to dark places to shed their wings until they grew a bit more and went wherever they went.

There were plenty of strange creatures in Jamaica that a prairie boy had never encountered in western Canada. The locals warned about scorpions and centipedes, which were the only harmful creatures. The early white inhabitants had decimated the snake population by importing the mongoose. But once the mongoose rid the island of snakes they killed all the chickens. We were told to place our footwear and clothes on a chair and to check them each morning for scorpions and centipedes. This lasted for about two weeks and then it was out of bed and on with the clothes with no thought of scorpions or centipedes. I only saw one scorpion and one centipede in my six months in Jamaica.

Sugar ants and cockroaches were extremely plentiful. Our cookies and sweets were provided in tins and if a lid were not fully closed the sugar ants would invade the area. The next person who dug into the tin for a treat was in for the surprise of his life. A number of ants about the size of hummingbirds crawled out of the can—not a pleasant experience.

I arrived about two weeks after the rest of the crew and the boys regaled me with stories of the number and size of the cockroaches. One evening before bedtime I turned out the lights and waited with a flashlight in hand. I had left a piece of steak on the dining room table. I could hear them rustling about. When I turned on the flashlight, the steak was nearly devoured and the wall, floor, and table were crawling with the biggest cockroaches I had ever seen. To keep them out of our stand-alone pantry, we placed a large water-filled can on the floor. Inside that can we placed the legs of the pantry in a smaller dry can. Obviously, if the insects were able to crawl up the water-filled can they would fall in and drown.

The Jamaican people were a resourceful group. Fresh water was in short supply and the government had laid surface water lines along all major roads, with a standpipe and attached faucet every one-half mile. There were also showerheads, and it was not unusual to see a worker stripped down to his birthday suit showering on the side of the road.

The Big Rig spudded in their first hole at an elevation of some thirty-five hundred feet and they had to have a source of water. They commandeered our slim-hole rig to dig a water well near sea level. The plan was to pump the water to the rig through a surface pipe. They determined the size of the pump, the length, diameter, and elevation change of the pipe, and therefore were able to calculate when the head of water should arrive at the rig. The expected time of arrival came and went with no sign of water. On investigation it was determined that the pump was okay, and upon further inspection we discovered that the locals had drilled holes in the pipe at different levels to obtain clear, sweet water. We posted guards and that solved the problem.

When we had spent the 3 million dollars and found no oil, we prepared to return to Canada. We held a large party for the ten white families at our end of the island. George Lynn, the shopkeeper from Sav-la-Mar, cooked a pit-fired suckling pig and we danced on the surface of the tennis court.

We then decided to hold a party for our coloured staff one Saturday shortly before we departed. We filled the four bathtubs with Red Stripe beer and ice and prepared oodles of food. The staff had a great time and we let our hair down and interrelated. Gordon, our gardener, came and advised that James wanted to see me in the living room. James was set up at one end of the room playing his guitar and singing western Canadian cowboy songs including those written by Wilf Carter. Apparently, he had visited the local library and read all he could about Calgary and Alberta. The tears were streaming down his cheeks and he implored me to take him to Canada as my servant. I explained that I would have great difficulty with Canadian Immigration and even if I solved that problem, I could not afford the wages in Canada. I told him how in Canada, only the very rich had servants. It was a tear-jerking moment and I still get teary eyed when I think of this special tribute. I will never forget James.

There was a wonderful bar and dining experience in the Blue Mountains above Kingston called the Blue Mountain Inn. They also rented rooms and when Nola and I returned in 1966 we stayed at this wonderful location. Luther and I visited there for a pleasant meal and became friendly with Mr. Black, the maître d'. We held our crew's farewell meal at Blue Mountain Inn before returning to Canada.

My time in Jamaica really opened my eyes to life in the larger world. Even though my experiences had not extended beyond western Canada, I had thought I knew a lot about the world. I was wrong.

I treated Jamaican people as I would anybody. I did not care if they were black, white, pink, or whatever and they treated me exactly the same. I had a wonderful experience with them. The only black people I had previously been in contact with were porters on the CPR and a school acquaintance.

When I arrived in Jamaica I saw how they lived. The abject

poverty of their situation made it look like aboriginal people in Canada were living like kings. I was really astounded. I was upset by how the majority of locals were living, but if you treated them properly, they paid it back tenfold. As a result I also learned not to prejudge people, countries, or societies until you have walked the same path. That was a major life lesson for me as a young man.

At that time, I was in the habit of making five-year goals, hopefully attaining a certain salary and position within that time frame. The Jamaican people never thought that way. Money did not seem to mean anything to them. They lived off the country. Some people said they were lazy, but that was not the case.

It was a whole different way of life for me, and it was a 180-degree change. My entire thinking turned around. Not all our people thought the same way. One of our drillers hated blacks. He did not know anything about them, but he hated them. And they knew it. In the movie *20,000 Leagues Under the Sea* there was a big black man who played a chieftain. Big Daddy was his name and he worked for us as a drill helper. One Sunday he came to me and he said, "Mr. Bob, do not send Mr. So-and-so to work tonight." "Why?" I asked. "Because he will not come back." Big Daddy had enough respect for me to warn me and we sent another driller out on that shift and shipped the errant employee back to Canada.

The Jamaican experience changed my attitude and I became a more sensitive human being.

A couple of days after I arrived I had attempted to E-log our first drill hole with a radioactive probe that was worth a fair piece of change. While drilling the hole we lost circulation and the drilling liquids disappeared into the cavernous limestone formation. We threw everything we could down the hole, including sugar cane hulls and all kinds of junk, to attempt to seal it off, to no avail.

The well was located on a sugar cane plantation owned by a black man. By two o'clock in the morning, I had been sweating on

An E-logging unit, Jamaica, 1956.

this problem for five or six hours, and the president of the company, Jack, said: "You better cut the cable." But I said, "Geez, I do not want to lose the cable and probe." "To hell with it," he said. "We'll get another one out of Florida … Cut the damn thing."

So I said, "Where do you want me to cut it?" "Anywhere you want." The next words to come out of my mouth were: "Eenie, meenie, minie, mo, catch a nigger by the toe." Then I realized what I had said, with this black man standing there watching the proceedings. Neither Jack nor the plantation owner commented on my indiscretion.

The plantation owner and I became good friends and I later apologized to him. "You do not have to apologize, Bob," he said.

"I know the saying." I was truly embarrassed and I learned a valuable lesson: be cognizant of your surroundings before opening your mouth.

The black Jamaicans persevered, with dignity and a laid-back outlook on life. They took you at face value and appreciated any niceties that you extended to them. I learned that they came from a background of oppression and slavery and that has had a continuing influence on them over the years. I returned to Canada having learned some humility and an understanding of people of colour, and people in general. I believe I became a much better person.

Bearing Gifts

On arrival back in Canada, in August 1956, my father, who was in Toronto, met the crew and me at Canada Customs. I had a couple of suitcases plumb full and a huge wicker basket loaded with gifts from Jamaica. The fellow we sent back because of his dislike for coloured people had written one of our drillers and advised we could bring back more rum than the normally limited one bottle. He also supplied the name of the customs shift supervisor who could be persuaded to take one brown and one white rum to turn his back on whatever excess rum a person was attempting to bring into Canada illegally.

I had the misfortune to be queried by a female customs inspector who, it seemed to me, disliked men, or at least me. When I unlocked one of my suitcases it flew open like a jack-in-the box. In the middle compartment were six bottles of rum: three brown and three white. She crawled up one side of me and down the other, berating me and advising I was only allowed one bottle of liquor. I asked if I could speak to the shift supervisor. He came over, took one brown and one white, and leaned over and opened the door of a cupboard and placed the two bottles with many others. He said to the customs lady: "Help the gentleman close up his luggage." This very plump inspector actually sat on my suitcase so I could close and lock it. I joined my father and the rest of the crew and we had a drink at the bar.

I flew to Regina where Nola and Hugh picked me up. They could not believe the amount of luggage I had. Off to Moose Jaw we went. I seldom wore a shirt under the Jamaican sun and I was just about as black as the typical Jamaican. I had purchased souvenirs for all and sundry: straw hats, straw place mats, straw and mahogany drink coasters, yards of colourful linen for Nola to make shirts for Brent and me, and dresses for her and Brenda, among other items. At that time I promised I would take Nola to Jamaica. She agreed and we set the date for ten years down the road, in 1966.

I took a month holiday and played golf nearly every day with the local businessmen at the Moose Jaw Golf and Country Club. On one occasion I offered them a drink of pure Jamaican Rum. They accepted and we met at our trailer. Normally, a one-half inch of pure Jamaican rum in a tumbler filled with your mix of choice was adequate. I attempted, with little luck, to convince them to use water as their mix as was the custom in Jamaica. They complained about the small amount of rum and I explained the potency of pure Jamaican rum. They insisted and I doubled the size. After one drink and much talking, etc., they stumbled out of the trailer in a very inebriated state. That white rum was powerful and to this day I can conjure up the rather pungent taste of Jamaican white rum.

While on this holiday, Nola, baby Brent, and I visited the Assiniboia bunch. We drank potent Jamaican rum and tequila brought back from Mexico by the local doctor. I became thoroughly zonked and at about 2 A.M. we headed the seventy-five miles back to Moose Jaw. Nola had never driven before and Norm Kyle gave her a quick driving lesson. We had to return, as Nola's mother was looking after Brenda and she had to be at work in the morning. Nola made it home with flying colours and the story has provided us with a lot of laughs whenever we have gotten together with the Assiniboia crowd.

Rintoul family upon Bob's return from Jamaica, 1956. Bob returned to Jamaica for a visit, with Nola, in 1966.

We returned to Calgary in the early part of September and parked and lived in the trailer at the Chinook Trailer Park. Northwest had been sold to Frank Hickey and Andy Anderson and they renamed the company Hickey Anderson Seismic Company or HASCO as it was better known. Jack MacMillan kept Luther and me on his personal payroll as there was another contract in the offing in Guatemala. That contract never materialized and Luther and I took one of our other slim-hole rigs, a 1500 Failing, and on Jack's urging went seeking contracts.

He said he would split the profits with us. We successfully picked up a contract with Western Decalta, which was headed up by Lindsay Richards, a rather crusty character. On our first visit, he looked at his watch and said, "Gentlemen, you have five minutes." I said, "Mr. Richards, thank you, but our proposal will take longer to present than five minutes," and I gathered up Luther and started for the door. Richards said, "Come back here, young men," and fifteen minutes later we had a contract for twenty holes in the Camrose area. I was learning that if you want respect you must earn it.

Luther was the driller, we hired a couple of helpers, and I looked after the surveying, E-logging, and managing the finances. We were reasonably successful and finished before the end of September. Although we received our regular wages, we never saw any of the profits from those contracts.

A Job with Accurate

I was still on Jack's private payroll when I had a call from Al Campbell, whom I had worked with in Assiniboia in the early fifties. Apparently, Accurate Geophysical had a three-year contract with Imperial Oil for three crews in the Whitecourt–Iosogun River valley area of north central Alberta and was short of party managers with bush experience. Accurate wanted to hire Al as one of their geophysicists but Al already had procured a job at Mobil Oil. They asked him if he knew of any party managers with bush experience and he recommended me.

I visited Bud Coote, president of Accurate, and he hired me on the spot. We moved the trailer and family to Edmonton and located it at the Skyline Trailer Park at 109th Street and Kingsway. Trailer lots were in short demand but we were able to secure a lot at Skyline due to our previous Chinook Trailer Park stay. Ron Southern of Atco Trailer, which was just a fledging organization in those days, owned both parks. The only lot available was the farthest north spot bordering 109th Street. It was located directly under the main east-west runway of the Edmonton Municipal Airport, and, at that time, the only airport in Edmonton. We bought our first car, a 1953 Studebaker Land Cruiser, which I used as a company vehicle and received mileage for company use.

The first chore I had as an employee of Accurate was establishing an office in Edmonton that was large enough to accommodate approximately twenty-five people. I found a satisfactory location on 104th Street near 70th Avenue. I took the first crew out and set up camp on the highway between Whitecourt and Valleyview on the south side of the Iosogun River valley. We moved in and started work in November. We worked until Christmas and had about fifteen days vacation. We drove to Moose Jaw to celebrate Christmas with Nola's parents.

After the holiday season I settled into camp with the crew and we proceeded to shoot some line. Imperial Oil had purchased one of the first sets of analog recording instruments and they were assigned to our crew. There were a lot of "bugs" in the system, so it was slow going. Because of instrument breakdowns, it was difficult to reach our monthly hour quota and the crew was pretty cranky.

Our rod man was shot and killed by an anxious elk hunter. Fortunately, the crew had not worked together long enough to form a bond or it could have been even more devastating. His parents visited the camp about a month later and asked to be shown where he had slept and was shot. I found that meeting very difficult. There was a trial in Edmonton for the hunter, who was charged with manslaughter. I was a reluctant witness. The hunter was exonerated when his lawyer called a renowned psychologist who claimed that after a two-week unsuccessful elk hunt he could easily mistake a partially visible yellow truck a half mile away for the rear end of an elk. The hunter was an older fellow and it was obvious that he would live with his horrible mistake for the rest of his life. I would never forget this man's face.

We were the feature crew for a showing of the first big successful track unit, the Nodwell. It provided me with an opportunity to meet many personnel from other oil and seismic companies.

In late February, Imperial Oil ordered us to move camp toward the Athabasca River. I chose a campsite and then hiked through the bush to where the drills were working. I instructed the dozers to clear out the campsite and doze a trail to the drills. This would allow us to return from our pending time off with little or no shut-down time. I contacted Worthing Transport, located in Edson, and requested that his trucks arrive at 7 A.M. on the day of our return.

The crew had departed for a well-deserved break and Dunc McLeod, our operator, and I were preparing to leave camp. The vice-president of Accurate showed up in a big huff demanding that I supervise the move on my time off. I carefully explained the arrangements I had made and that we would not lose one day's work. He told me that he had flown the area and that my plan was impossible. I reminded him that I had walked the route and it could be done. He said, "Well, if you won't stay, how about Dunc?" "No, he deserves his time off," I responded. Then he suggested that the surveyor remain behind. "Nope, he's left." In my mind there was no need for further conversation so Dunc and I loaded up the car with the day's records and headed for Edmonton, leaving an incredulous vice-president standing in the middle of the camp.

We returned the first of March and my plan worked like clockwork. The day of the move the cook had a hot meal on the table for the crew when they returned from the field. I knew my days were numbered with Accurate because of my "run in" with the VP. Also, I had plans of becoming a field supervisor and they had hired a fellow with considerably more experience than me, so I knew I was at the end of my trail as far as promotions were concerned.

Opportunity Knocks

We were buying our explosives and drilling mud from Continental Explosives out of Valleyview. I dropped by their office one day and had a long discussion with their owner and president, Dick Strazer. We talked about my experiences in Jamaica including the necessity of ordering supplies two months in advance. He expressed a desire to expand to Australia to supply explosives to the Snowy Mountain Hydro project.

About two weeks later he arrived at camp and wanted to have a private talk. He offered me a "drink" and I advised him we operated a dry camp. We drove down the road a couple of miles, parked, had a drink, and he offered me a job as sales manager with the same salary as I was making, a membership at the Edmonton Petroleum Club, a golf course of my choice, a car, and an expense account.

He suggested opening an office in Edmonton, which made sense as we were living there. He also explained that Continental had a very limited line of credit and he was trying to sell the company. He assured me I could follow him on his next venture or go with the purchasing company. I told him I would talk it over with my wife and let him know. Nola agreed and I gave him an affirmative answer. I drove to Edmonton and met with the president of Accurate, Bud Coote. He was very disappointed that I was leaving and I pointed out my "run in" with the VP and their hiring of a field supervisor. He reluctantly agreed to my resignation and I departed at the end of March.

April was an exciting month for me, what with commencing my employment with Continental and setting up an office on Whyte Avenue across from the old south side Edmonton Post

Office that had originally served Strathcona (Edmonton, south of the Saskatchewan River, was called Strathcona before amalgamation with the city of Edmonton in 1912).

The excitement continued as Continental supplied a 1957 Pontiac, standard stick, two-door car for twenty-four hundred dollars. I also joined the Edmonton Petroleum Club located in the old Airliner Hotel near the airport on Kingsway Avenue and the Edmonton Golf and Country Club. Nola and I continued to live in our trailer at the Skyline Trailer Park.

My first endeavour as a sales manager and salesman was to attempt to secure some explosives and accessory business on the construction of the TransCanada Highway near Banff. It became immediately apparent that my knowledge of explosives was very limited. This would not do so I read the CIL (the manufacturer) Explosives Handbook in one night. As it turned out the explosives and accessory business contract had been let.

Dick Strazer was an interesting fellow. He and I spent a lot of time together to acquaint each other with our backgrounds and general information. Dick was born and raised in Chicago, where he married his wife, Shirley. They moved to Seattle and Shirley worked at the famous Ivar's Fish Bar while Dick busied himself as a beachcomber. They remained there for about a year when Dick's wanderlust called and they took off for Alaska, where Dick planned to attend university. The original route linking Alberta highways to the Alaska Highway wound from Edmonton up Highway 2 to Athabasca or jumped off on Highway 44 where Highways 2 and 44 joined at Lesser Slave Lake. From there, the road wound around Lesser Slave Lake and through High Prairie to Valleyview. When their vehicle broke down between High Prairie and Valleyview they did not have the money for repairs. They discovered an abandoned roadhouse near their breakdown and finagled some help and money, fixed it up, and reopened the restaurant. They remained there about a year and were reasonably successful.

By now Dick had purchased another vehicle and they decided to raise chickens and sell the eggs in Alaska, where they sold for an exorbitant price. Dick mounted a heated camper unit on the truck and started off on his first trip, leaving Shirley to run the restaurant. Partway up the Alaska Highway the propane heater in the camper unit exploded, destroying all his eggs and the truck. Dick was badly burned.

During his recuperation Dick received an offer from a fellow in Valleyview by the name of Reber, who owned the BA (forerunner of Gulf Oil) bulk plant, the power plant (in the early days of small towns in northern Alberta and Saskatchewan, the power plants were privately owned and only operated daily from 6 A.M. to 10 P.M.), and the telephone system (same applied as with the power plants). Reber offered Dick and Shirley the opportunity to work for him for a share of the profits. They jumped at the opportunity.

While delivering fuel to Century Geophysical, the supervisor asked Dick to start an explosives distributing business and told him that they would purchase all their requirements from Dick. In the early days of oil exploration the majority of seismic crews were fresh out of the U.S. They were accustomed to ordering explosives and accessories up to midnight and expected delivery by 7 A.M. the next morning. Currently, they had to order directly from CIL, located in Edmonton, two weeks in advance. The supervisor offered Dick a fifty-cent per hundred weight premium over CIL prices. Explosives were selling at that time for about fifteen dollars per hundred weight.

Dick and Reber formed Continental Explosives Limited, ordered a couple of trucks specially outfitted to haul explosives, and away they went. CIL reluctantly sold to Continental only because they were cognizant of the Canadian Government anti-competition regulations.

In 1955 the United States passed anti-combine legislation ordering DuPont to divest their GMC and CIL shares. They owned CIL jointly with Imperial Chemical Co. (ICI), headquartered in Britain. That series of events allowed ICI to gain sole ownership of CIL. However, CIL did not make it easy for distributors, as they wanted all the gravy, including trucking.

There were two other distributors in Alberta, Webb Distributors in Drumheller and Black Hardware in Lethbridge, who with the blessing of CIL supplied small quantities of explosives and accessories to the coal mining industry. However, with the advent of seismic exploration those companies were allowed to take advantage of this new opportunity.

Dick built the business over two years to include branches in Valleyview, Peace River, and Fort St. John. He purchased Reber's interest in Continental. As pointed out earlier, there were over fifty crews operating or supplied out of Peace River. But his fifty-eight-thousand-dollar operating line of credit was inadequate, particularly when most of the crews paid in ninety days.

Dick loved flying and acquired a wheel-float Cessna 182 plane. An Edmonton broker administered the insurance. Their employee, Hal Comish, handled Dick's insurance. Hal knew everybody there was to know in the Edmonton business world from the federal member of parliament to the heads of large companies.

During 1957 construction commenced on the TransCanada and B.C. Interior Pipelines. Perini was the contractor on the first section of the TransCanada Pipeline requiring blasting east of Winnipeg. Hal introduced Dick and me to Bill Aldridge, vice-president of Perini, and he suggested we contact their spread boss, Rusty Killingsworth, at Ste. Anne, Manitoba.

Dick promised me that if I secured this and the Inland job, I could play golf for the rest of the summer—a pretty amazing offer. However, the summer of golf did not materialize, thanks to my personality.

I flew to Winnipeg, picked up a U-drive, and drove to Ste. Anne, located about fifty miles east of Winnipeg. I met with Rusty Killingsworth, the spread superintendent, and offered him our services at a premium over his purchase price from CIL. CIL had a packaging plant at Brainerd, Manitoba, so distance was not a factor. I offered a large explosives magazine (for storage) located near their job. He retorted that he had located an abandoned magazine used when the TransCanada Highway was built. I had previously determined that this magazine was in poor condition and too close to the highway. He would not budge from his position so I suggested we would not only provide an up-to-date main magazine but would also provide a number of seismic explosives magazines (portable, with a capacity of four thousand pounds). I suggested he could remove the wheels and place skids under the magazines for ease of transportation over the muskeg. We would provide a magazine operator to fill the portable magazines daily for their next day requirements. They could skid the magazine to the job site, returning the magazine to the nearest point of truck availability. The terrain was very rugged, Precambrian Shield overlaid with about three feet of muskeg. I advised him this would cost him three dollars per hundred weight over the CIL quoted price. He grumbled about the extra cost and dismissed me. I drove back to Winnipeg and caught a plane back to Edmonton.

The next day while I was making customer calls in Edmonton Nola let me know by radio phone that Rusty wanted me to call him. When I contacted him, he asked, "What's the deal?" I reiterated the offer. He said, "Get your ass down here and see me in camp at 7 A.M. tomorrow morning." It was about 3 P.M. and I caught the last plane out of Edmonton for Winnipeg. I rented a car, located a hotel room, and rose at 5:30 A.M. the next day and drove to Ste. Anne.

It was raining and had rained all night. The campsite was a sea of mud and the nearby river was overflowing its banks. I went to the office and the office manager directed me to Rusty's house trailer. When his wife opened the door I was met by two Dobermans and ushered into the trailer. Rusty asked me to explain my offer, yet again. "You have a deal," he said, and we shook hands. With that he pulled a case of Johnny Walker black label whisky from under the bed. "You leave when the bottle is finished."

About two hours later I stumbled to my car and drove a couple of miles up the highway, pulled onto the shoulder, and slept for three hours. Rusty Killingsworth was a real character. He was a big man, with a bald head, except for a rusty fringe around the lower skull. On this job when he became unhappy with the progress of the pipe layers, he would go into a rage and start pounding his head on the pipe string that was about to be lowered into the ditch.

When I arrived back in Edmonton I arranged to send three seismic explosives magazines, a truck, and Bert Allan to Ste. Anne. Bert built the main magazine and we were in business. CIL was extremely unhappy about us taking this job but it made them realize that we were for real. As DuPont was moving into Canada to give CIL their first taste of competition, a thread of fear and trepidation ran throughout CIL. They offered us the opportunity to join forces with them. We would receive a percentage on all successful pipeline bids to the Lakehead.

CIL's manager in Winnipeg had never dealt with the oil industry and their demands. He asked to accompany me on a sales trip. He did not make a very good impression on that first call, and the spread superintendent told me to leave him at home the next time I visited.

Since the U.S. Anti-combines Commission had ordered DuPont to divest its interest in CIL, the TransCanada Pipeline opened the door for DuPont to invest in an explosives manufacturing plant in Canada. DuPont commenced manufacturing explosives in their North Bay plant and were giving CIL a run for its money on the TransCanada Pipeline. Up to that time CIL had carte blanche on all explosives business in Canada, and the company became completely paranoid concerning competition from DuPont. That attitude played into the distributor's hands.

A Wild Goose

With a couple of spreads in our pocket on the TransCanada Pipeline, I then focused my attention on the BC Inland Line. Dutton Williams was the contractor on this job. The line's route represented a horseshoe connecting Salmon Arm, Vernon, Kelowna, Penticton, Grand Forks, Rossland, Trail, Castlegar, Nelson, and Salmo. As you can imagine there was a lot of rock to blast.

Dutton Williams's main office was in Penticton and contacting the managing superintendent was impossible. The spread bosses were busy from five in the morning until midnight. I arrived after construction commenced and CIL was supplying their explosives and accessories from its regional office and explosives magazines in Nelson. Delivery problems were prevalent due to long delivery distances and near impassable mountain trails.

The spread bosses met every morning at 5 A.M. at a location near Penticton. They staked a large circle with wire rope and those who wanted a job stood inside the circle and the spread bosses picked out which employees they needed for that day's work. I attended this show for a couple of mornings without successfully talking to any one of the spread bosses.

On the third morning, an older gentleman came up to me and told me that he had been observing me the last two mornings and as I did not appear to want a job, what did I want? I told him my story and explained about the success of the seismic magazines and arrangement on the Perini section of the TransCanada Pipeline. He showed some interest and suggested I contact their purchasing agent in Penticton. He felt I had a plausible solution to the delivery problems. It turned out that the gentleman was Guy Williams, who started and was president of Williams Brothers Pipe Line Company.

That same day I made arrangements to visit the purchasing agent, a Mr. Scoffins, whom I found out, through discussion, had gone to school with Nola in Moose Jaw. I presented my proposal to him with the three-dollar up charge per one hundred pounds. He accepted and I went to my motel and prepared a written document for his files. We were in business.

It was necessary to transport the explosives from CIL's closest manufacturing facility at James Island, an isolated location a little southeast of Pat Bay on Vancouver Island. The accessories were available in the interior of BC at Tappen (off the TransCanada Highway just west of Salmon Arm), where CIL had a detonator and prima cord manufacturing and storage facility.

The explosives were transported by rail car on barges to the mainland and either trucked or sent by rail to the BC interior. What with attempting to accommodate tides from the island to the barges, this system was very time consuming and cumbersome. The delivery system did not lend itself to properly supplying a high production industry such as pipeline construction.

We purchased a USA Navy LCVP landing craft from a war surplus supplier near Seattle. LCVP stands for Landing Craft-Vehicle, Personnel. These barges were used for beach landings during World War II. They ferried troops and equipment from the ships to the beaches during the invasions of France. They were flat bottomed, about sixty feet long by twenty feet wide, and were powered by a 670 horsepower GMC diesel motor. They had a wheel for navigating the craft and one end dropped down for easy entry and exit. The entire craft was open to the elements.

We needed a name for the craft. I thought the whole thing was a wild idea, so I suggested "Wild Goose." The name stuck. We were anxious to have it fitted out in Vancouver for transporting explosives and time was of the essence. We loaded a couple of sleeping bags and a cooler full of food and gingerly motored up the coast after dark as close to the shore as possible. We circumvented customs and the naval authorities, and after three harrowing nights we docked at a prearranged ship refitter.

The shipwrights spent an interminable amount of time outfitting the craft with a wheelhouse complete with bunks, a small galley, ship-to-shore phone, and a depth sounder. They installed a snugly fitted tarpaulin to cover the storage area. The front ramp was welded shut. The running gear, engine, etc., were tested and upgraded where required. It took two weeks to complete the work.

In the meantime we shipped the seismic magazines to the pipeline sites, set up a magazine operator, and delivered explosives using the transportation system already in place.

Dick and I contacted the West Coast unions responsible for supplying ship crew members and attempted to negotiate a contract. They demanded at least six crew on this little craft and a plethora of conditions we could not deal with physically or financially. We decided to go it ourselves. We had a truck driver, Audie Nordstrock, on staff in Fort St. John who held a Finnish ocean craft captain's licence. We summoned him from Fort St. John and within a couple of days he was issued his Canadian captain's licence.

On a bright, warm Saturday afternoon we were informed that the Wild Goose was ready to sail and we shoved off from the shipbuilder's jetty and headed across the Straights of Georgia. Although the weather was clear and warm there were six-foot swells running and in a flat bottom boat it was a rather rough trip. After about half an hour we were listing badly to port. After investigating the hold we determined that we were shipping water. Our erstwhile shipbuilders had not cleared the bilges of debris from the renovations. While Audie captained the craft, Dick and I lowered ourselves into the hold to clear the bilge pump inlets of material. It was about 70 degrees Fahrenheit and with the canvas covering the hold the temperature was close to 100 degrees in the bilge. What with the smell of diesel fuel and the heat we came very close to flipping our biscuits. After clearing the pump intakes we attempted to use the electric pump assist to no avail so we were required to man the hand pumps.

We returned to the ship builders and they completed the job of cleaning the refuse from the bilges and repaired the electric bilge pumps. We shoved off the second time, around 4 P.M. We contacted the James Island plant to warn them of our late arrival. They arranged to leave a key for the storage magazine. We loaded twenty thousand pounds of explosives as fast as possible and headed up Active Pass—the main route for sea traffic to and from Pat Bay.

Audie started out captaining the boat but declared he was tired and crawled into the top bunk. Dick took over and about a half-hour later declared his intention to crawl into the lower bunk. That left me to captain the craft through Active Pass. This route is named Active Pass because it is very narrow and deep. To keep in line with the buoys (navigational markers), it is sometimes necessary to crowd the shoreline. The depth of water changes rapidly so you have to be very cognizant of the depth sounder readings.

Where the pass empties into the Strait of Georgia calls for a very deft maneuver; the deep-water route requires a 135-degree turn to starboard and the tides are furious. As we made our starboard move, a large Black Ball Ferry (the predecessor to BC Ferries) was bearing down on us. We would be like a pimple on a gnat's ass to the ferry captain. I was successful in escaping that predicament and headed for the Fraser River inlet using the Vancouver airport marker lights for direction. At about 2 A.M., we were running dangerously low on fuel so we tied up to a floating Esso fueling station until they opened at 7 A.M.

After fueling we carried on to the Fraser River and tied up at a jetty where two of our trucks were waiting. We quickly loaded the trucks and they headed for the interior. One last comical incident took place when we laid up at the jetty. We were so anxious to load the trucks that we forgot to take into consideration the out-going tide. We returned to the Wild Goose only to find it high and dry.

We eventually added another LCVP, sans engine, to our fleet and pushed it in front of the Wild Goose. This allowed us to transport forty thousand pounds of explosives per sailing. We named the second craft Wild Goose II.

Dick and I looked at supplying the Ripple Rock job. This was a very large project aimed at removing the crest of an underwater mountain to enhance shipping on the Inside Passage from Vancouver to Alaska. The Inside Passage runs north and south between the mainland and Vancouver Island. The Ripple Rock waters were very difficult to navigate and many ships had sunk attempting this route. The Inside Passage cut off days of travel and the weather was calmer. The alternative was to navigate the open waters west of Vancouver Island.

The engineers had determined that the most cost-efficient way to remove the cone of rock was to use the "coyote" system of blasting, where tunnels were blasted into the dome on a predetermined pattern and then loaded with explosives. It was determined that 1,375 tons of explosives would be required. It was the largest non-nuclear blast in the world up to that date.

We were not the successful bidder on the contract. DuPont got the job. This was a major blow to CIL and another rung in the ladder for the recently established DuPont Canadian explosives manufacturing plant. A number of seismic recording stations around the world, including in western Canada, recorded the blast.

As stated previously, after procuring the two pipeline jobs, Dick had promised I could play golf for the remainder of the summer. As tempting as this plan was it was not in my makeup to be unproductive. It became very clear that Continental Explosives's

Bob with a coveted "Interpreter" trophy from Doodlebug Golf Tournament, 1980. Golf was an important part of Bob's life from the day his parents introduced him to the game at age twelve.

accounting practices were suspect, so I asked Dick if I could bring the accounting up to snuff. He agreed. I spent the remainder of the summer completing this task as well as visiting and securing more seismic business in the Edmonton area. I also travelled to the different branches, counselling them in accounting and sales.

That fall Nola, Dick, Shirley (the owner's wife), and I attended the Doodlebug Golf Tournament. Since 1953 the tournament has been a yearly get-together of contractors and oil company and supplier personnel active in the geophysical industry. Due to my years in the geophysical business, it irked me that up to now I had been unable to attend the "Doodlebug." This was due to commitments as a party manager. The tournament is held at the CPR Banff Springs Hotel Golf Course (now called The Fairmont Banff Springs Hotel and Golf Course). I played in the Doodlebug until 1983 and won my flight in 1962, 1977 (twenty-fifth anniversary), 1978, and 1980. All but the twenty-fifth anniversary trophy, which was a sterling silver tray, were ceramic caricatures of such seismic entities as interpreters or party managers. The ceramics were and are a most coveted trophy.

Dick had discussions with Gordon Black, president and owner of Explosives Limited (XL), the CIL distributor for southern Alberta, about purchasing Continental. The discussion culminated in Banff at the Doodlebug. Dick gave me the option of continuing with him in a hunting-guiding operation out of Watson Lake, Yukon, then owned by Dal Dalzell, or joining Explosives Limited. His new venture consisted of an office in Watson Lake, a float plane, and a couple of lodges in the mountains. The area was known for its record Dall sheep and attracted a large number of American hunters. Although I liked Dick very much, I felt there were better opportunities for my family in joining Explosives Limited.

Explosives Limited

XL maintained a small office on Macleod Trail in Calgary. Black engaged Don Telfer as general manager of the extended corporation. Don's prior employment was as western manager of Liquid Air, a large welding gas and supply company. Black's hardware company, Western Canada Hardware, headquartered in Lethbridge, was a distributor for southern Alberta and western BC for Liquid Air. I was appointed "general everything" and continued in the sales and operations end of the business.

During our first family holiday since Nola and I were married, we travelled in the summer of 1958 to Ottawa to visit my parents. Don Telfer had agreed that I was deserving of some time off. Holidays, unlike today, were not automatic. My dad and mom took us to New York City and on the way we stopped at the "North Pole." This was an amusement centre in northern New York State that emulated the North Pole with Santa Claus, reindeer, elves, workshops, and other animals. Brent and Brenda loved the show.

We also stopped at the Lake Placid 1932 Olympic site because of my interest in sports, especially hockey. While in New York City we took the elevator to the top of the Empire State Building and the Statue of Liberty and enjoyed a boat trip around the Island of Manhattan. I had accomplished two of my four wishes: to see the Empire State Building and the Statue of Liberty. The other two wishes, San Francisco and New Orleans, would be realized a number of years later. The holiday was a good change of pace for me and I felt ready to take on the challenges of a new job.

The town of Inuvik, located on the east shore of the Mackenzie River basin, had been built by the federal government in the 1950s, ostensibly to replace Aklavik on the west side of the river. Since a Hudson's Bay Trading Post was established in the region in 1912 Aklavik had been slowly sinking into the permafrost. Inuvik would be the new Aklavik. A large percentage of the Northwest Territories is composed of Precambrian Shield overlaid with a thick layer of permafrost. Inuvik's water and sewer systems were delivered by utilidors positioned on supports about three feet above the surface of the permafrost. The construction of the utilidor is an ingenious method of two small pipes: one for water, the other for sewage, encased in a larger insulated heated pipe.

The Canadian government spent 40 million dollars in 1959 (which correlates to over one-third billion in today's dollars) to build the new town. The original tender did not attract a single bidder due to the supply distances and the unstable ground conditions. Finally, the federal government accepted a bid from a consortium of the largest contractors in North America with Mannix as the operator.

One day, Bob Lindeman joined me on the golf course. It turned out that he was the purchasing agent for Mannix's Inuvik contract. CIL was directly supplying explosives and accessories for blasting rip rap rock (broken rock compacted as a base for airport runways). The explosives were currently being trucked to Hay River, located on Great Slave Lake, and forwarded by barge down the Mackenzie River to Inuvik. From discussions with Lindeman I assumed they were experiencing logistical problems relative to delivery. The alternative, if they missed the barge season, was very costly air transport.

If he was willing to change his explosives supplier, I suggested that we could take care of their delivery problems. We would establish an explosives magazine complete with a year-round back-up inventory in Hay River, and an agent. This method would guarantee them a supply of explosives and accessories. Bob awarded us the contract.

I frequently travelled by vehicle to Hay River as an extension of my northern branch visits. Hay River was seven hundred miles of 90 percent gravel roads from Edmonton. Our son Brent, when he was five years old, joined me on a trip to Hay River. What an experience for him, viewing the Alexander Falls on the Mackenzie, Native communities, and temperatures down to minus 70 Fahrenheit. He still remembers this trip, fifty years later.

During the summer of 1959 this contract afforded me the opportunity to travel by boat down the Mackenzie River. Yellowknife Transport, our barge contractor, competed against the federal government's barge operation, Northern Transportation. The owner of Yellowknife was Earl Harcourt and he offered to take me down the Mackenzie on his private boat, a converted PT craft from World War II, powered by two 670 Jimmy Diesel motors.

The boat had a cabin containing a galley (kitchen), head

(bathroom), and sleeping accommodations for two. The crew consisted of Earl, his wife (who cooked for us), and his general manager. Behind the bridge and over the motor compartment was a covered area where the general manager and I slept in sleeping bags. It was August and the northern lights were in full display. What a sight that far north! From Hay River to Inuvik was more than one thousand miles and it took us two weeks, running with the current. However, we did assist a number of tugs in navigating against the flow on the numerous rapids.

Yellowknife Transport had the largest tug on the river with a 1,280 horsepower motor. The tugs pushed huge barges, as many as five at a time, loaded with all the products required: aviation fuel, motor vehicle fuel, heating oil, diesel fuel, groceries, frozen meat, etc. The supplies were off loaded in large enough quantities to allow the various river communities to survive the long winter months.

We visited all the communities along the way: Fort Providence, Fort Simpson, Wrigley, Norman Wells, Fort Good Hope, and Fort McPherson. Before docking at Inuvik we travelled an additional ninety-three miles to Tuktoyaktuk, on the Arctic Ocean. I disembarked at Inuvik and visited the general superintendent of Mannix. A large quantity of explosives was used to blast rip rap from the Precambrian Shield to build the airport runway.

While in Inuvik there was an Accurate geophysical crew carrying out seismic exploration on the river and lo and behold they were using the old Continental Explosives landing crafts, the Wild Goose I and II. It brought back fond memories from my west coast days when we originally commissioned the two craft. The Inuvik jetty produced trout weighing up to twenty pounds, which we caught on a hand line baited with a piece of bread!

While in Inuvik I also visited a Shell Oil seismic exploration crew. They commandeered me to accompany them on their plane to Aklavik. As Inuvik was still in the early stages of completion, liquor was at a premium. Booze was available in Aklavik, but only with a government issued permit, as it was rationed. I represented an additional permit holder and an added supply of liquor. It afforded me an opportunity to view Aklavik before it was downgraded once Inuvik was established.

Bob Lindeman came to Edmonton from Australia. He had been the British attaché to Chiang Kai-shek, then ruler of China. He was also known to Chou En-Lai, the Communist Chinese leader who took over China in 1949. The day before Chou En-Lai did so, he advised Bob to take his family and flee, as Chou could not guarantee his family's safety. Bob also described numerous stories about Madam Chiang Kai-shek, who Bob says was absolutely ruthless. Bob was a grandson of the founder of Lindeman Australian wines and as a side business was instrumental in importing the first Lindeman wines into Canada for resale.

I travelled numerous times to visit gold mines in Yellowknife and uranium mines in northern Saskatchewan and the Territories. Conditions in Yellowknife in those days were very primitive. During the winter the taxis never shut off their motors as the temperature rarely rose above minus 40 degrees Fahrenheit. With two gold mines in operation, the bars remained open twenty-three hours a day. The extra hour was used to sweep out and lay new sawdust on the floor. The ravens were the largest I had ever encountered and got their jollies from playing tag with the local sled dogs as they chewed bones.

Within a year Explosives Limited hired a CIL employee, Stan Graham, who was appointed sales manager. I took on the operations manager's position. Other than Central Alberta, where Tommy Webb operated Webb Distributors, XL had no competition. In 1959 the company advised me to buy a home in Edmonton, as I would be there for a long time. We sold our trailer to Fred Zouboules (I had worked with him in Jamaica) and his family. Fred was our branch manager in Fort St. John. We accepted some of his furniture as part of the overall sale. We bought a home in the Hardisty-Goldbar area of Edmonton, 10811 Forty-fourth Street. I moved our office from Eighty-second Avenue to south of Edmonton on the Calgary Trail (Highway 2). Leduc Construction owned the building and occupied the lower floor. This company represented the start of the present construction giant Ledcor.

In the meantime XL built a new office in Calgary on Blackfoot Trail and in 1960 asked us to move to Calgary. So much for a long tenure in Edmonton.

Haysboro Days

When we purchased our first home in Edmonton in 1959, we were required to put 25 percent of the total house value down and take out a mortgage over twenty-five years. Houses were difficult to sell in 1960, due to a change in mortgage regulations, where only six hundred to eight hundred dollars were required as a down payment with the remainder over twenty-five years. There were a number of oil companies shifting personnel back and forth between Calgary and Edmonton and I came up with "why not trade?"—no real estate fees. I placed an ad in the Calgary *Herald* and within days we completed a trade with a Texaco employee transferred to Edmonton. The two homes were within twenty-five square feet in size and twenty-five dollars in value.

We moved into our new home in the Haysboro district of Calgary on July 1, 1960, and moved to Strathmore in 1979 when we sold the house to our son, Brent, and his wife, Carla. The Haysboro house was ideally located, as primary, junior, and senior high schools were within easy walking distance. The address was 8708 Eleventh Street SW, and was a ten-minute drive from work.

We were quite active in community affairs in Haysboro. I coached bantam hockey, assisted by Doug Cooper, Nola's brother, also a Haysboro resident. Brent played hockey and progressed through the different divisions. In 1967 he was a member of the

Haysboro Pee Wee B team, which won the City Championship. Brenda was runner-up at the annual Winter Carnival Snow Queen contest in 1963 and was elected Snow Queen in 1967. Nola helped design the kitchen when the new community hall was built. We saw construction of the south side "Y," Heritage Park, Rose Kohn Arena, and the Rockyview Hospital while living in Haysboro.

My position remained the same, operations manager. However, being located in the head office afforded me the opportunity to advance my business experience. I was appointed second-in-command and my primary duties were to make certain our operations ran efficiently and within budget. I also contacted seismic contractors and oil company geophysicists on a continuing basis. My knowledge of the seismic business and personnel was superior to others in the company, due mainly to my past employment in the industry.

I travelled, at least twice a year, spring and late fall, to all our seismic supply branches to ensure we were ready for the winter rush and were in clean-up mode for the summer. I also travelled to our construction and mining branches a couple of times a year and called on mining and construction offices to keep up contacts and procure new business. It was amazing to me how many of my old school acquaintances were in positions of authority in the mining and construction business. Telfer and Graham also made similar trips during the year. We were very successful and working with Telfer allowed me to become well versed in extended business practices. The business knowledge I absorbed at Explosives Limited was priceless and was far more extensive than the training I would have received had I completed a university MBA.

We eventually set up permanent branches in Calgary, Medicine Hat, Pincher Creek, Lethbridge, Edmonton, Valleyview, Peace River, High Level, Hay River, Inuvik, Fort St. John, Fort Nelson, and Whitehorse. We supplied explosives to nearly every mountain road construction and mining site in our territory: southern and northern Alberta, western BC, Yukon, and the Northwest Territories.

During the summer of 1961 I travelled extensively to and through the Crowsnest Pass and beyond. My first trip was in the early spring of 1961, and at that time they were still mining coal underground. About ten miles inside the border of British Columbia there were strings of coking ovens parallel to the highway. I arrived at this location on a very rainy spring evening. The fumes and smoke from the ovens reminded me of stories I had read about manufacturing in 1800s' Europe. It was surreal, like descriptions of Hell.

During that first trip, I scouted a planned pipeline in very mountainous country, south and east of Elko (about forty miles west of Fernie on Highway 3). As I proceeded along this barely visible trail the topography became extremely difficult and I finally bottomed out and tore a hole in the oil pan of the vehicle. I walked back approximately ten miles to Elko and contacted the nearest tow truck in Fernie.

The tow truck finally arrived around 5 P.M. and by the time we located the car it was getting dark. The tow truck operator took one look and said, "If I'd known where your car was I would not have left Fernie." He refused to go to the bottom of the valley where the vehicle was stranded but insisted I fill the crankcase with oil and drive it to the summit where his tow truck was located. We did not arrive in Fernie until 10 P.M. In those days Fernie was a ghost town. All the underground coal mines in the area had been closed for some years and there was only one hotel open. The next day on dismantling the motor of the car we determined that the vehicle required a new crankshaft and bearings. This meant a further delay, as the parts would have to come from Calgary.

It was a weekend and I called Nola to let her know of my predicament. She had undergone a scary night as a prowler had attempted to enter the house. There had been a rapist whose *modus operandi* fit this situation, so rather than remaining in Fernie I took the next bus home to Calgary. We never had any more trouble with the culprit but we did buy a very large German Shepherd. Eventually, the police caught this criminal. He had worked for Keith Construction, the contractor who built most of the homes in Haysboro. He knew the layout of the homes and would case them to make certain the man of the house was away and there were no dogs. He would park his vehicle in an alley some distance from the intended victim's house and if unsuccessful in breaking into the home would run to his car and disappear. He attacked a number of women and it took the police over three years to apprehend him. I took the bus back to Fernie and picked up the car the following Wednesday.

We were fortunate in being awarded the contract to supply explosive and ancillary products to the W.A.C. Bennett dam at Hudson Hope, BC, on the Peace River. It was a gargantuan project and afforded me the opportunity to expand my knowledge of the construction industry.

We were on the job from pilot tunnel to blasting the plug. The pilot tunnels were blasted to prove up the type of rock they would encounter in the actual diversion tunnels. The plug was an area of rock left in place to dam the river and allow the water to flow through the diversion tunnels thereby protecting the construction site.

When the dam, pen stocks, and turbines were completed the plug was blasted to allow the water to enter the pen stocks. As a matter of interest when I took my mother to Dawson City for the 100th anniversary of the Klondike gold find, we toured the dam site. While in the turbine area I was able to explain to the guide why you could see half of each drill hole in the rock walls. This, I explained, is evidence of a first-class blasting job.

In 1965–66 we were also involved in a joint operation with Continental Explosives (1958 Ltd.) to supply explosives and ancillary products to the Kleenley Side Dam on the Columbia River near Revelstoke, BC. We were the operators and I worked hand in hand with the Continental representative, Ken Kidder. He was the father of Margot Kidder, who starred as Lois Lane in the movie *Superman*. I met her when she was a little girl living in North Vancouver.

This was similar to the W.A.C. Bennett job, but we were also required to inventory the job so we hired an independent accountant to work with me. It turned out he was an old high school associate of mine. He was having trouble figuring out how we were going to tackle a particularly thorny problem. I slept on it overnight and came up with a solution. His comment was: "I did not realize you were an accountant."

The Reunion

On April 15, 1966, Nola and I left for our promised Jamaica trip. This was our first major trip as a married couple without the children. Aunt Jean agreed to babysit the children and, in passing, mentioned that Laura, my cousin, and her husband, Walter, would be vacationing in Mexico City. I contacted them in Hollywood and arranged to meet them in Mexico City. For the last ten years Nola had taken on jobs including ironing and office work for neighbours to provide money for a very fancy negligee to surprise me in Jamaica. This trip was essentially our honeymoon because we could not afford one when we got married. While planning our trip, I found out we could travel from Calgary to Mexico City to Jamaica and return through New Orleans and Houston to Calgary for very little more than a direct return flight to Jamaica. It snowed the day we left Calgary, necessitating de-icing of the plane before take off. We arrived many hours late into Mexico City and when we attempted to register at the hotel with our guaranteed reservation, we were told they did not have a room for us. The clerk also claimed he had no record of our reservation and when I showed him my written record, he shrugged his shoulders and feigned ignorance of the English language. I was furious and with my lack of Spanish and his lack of English, the situation was getting out of hand. Fortunately, a Mexican gentleman came forth who spoke impeccable English and procured a room for us within thirty seconds. It turned out he was a doctor and had travelled to Canada many times. We were very grateful for his help.

Our Mexico City experiences were priceless. We arose about 9 A.M., ate breakfast, and walked around the city, which at that time had 12 million residents. We favoured Juarez Avenue, and ate lunch there daily at about 2 or 3 in the afternoon, entering whichever cafe was nearby. We visited Xochimilco floating gardens, Maximilian's Castle (he was an emancipator of Mexico), Chapultepec Park, Zócalo (the city square where the Mexican Government sits), and the Teotihuacan Pyramids of the Sun and the Moon. We also attended the Folklorico Ballet at the Palace of Fine Arts. After the performance we dined in the penthouse dining room of the Latino Americana Hotel.

Laura and Walter joined us for a couple of meals and loaned us their rental car to go to a bullfight outside of Mexico City. Apparently, we hit the only Sunday of the year that bullfights were not held in the city. We became hopelessly lost and had to hire a taxi to lead us back to the hotel.

I had one scary experience in Mexico City. Each night after dinner we stopped at a nearby bar for a beer. This particular evening, Nola decided to go directly to our hotel. While sitting at the bar a gentleman sat down beside me and struck up a conversation. He asked where I lived and I had to draw him a map showing Calgary, very nearly straight north of Mexico City. He offered to buy me a drink, which I declined and he immediately became very irritated, requesting to see my passport. I also declined this request and he became belligerent. The bartender showed up and talked to the chap and while doing so suggested I leave, which I did. The next day I revisited the bar and the bartender told me that the individual was a Mexico City policeman with a drinking problem and would probably have arrested me when I could not show my passport because Nola had it in her purse. After a few days we continued to Jamaica.

We arrived in Kingston, the capital, on a Sunday and stayed at the Blue Mountain Inn. On the advice of our bank manager, before we left Calgary we had mailed a bank draft to the Bank of Montreal's affiliate in Kingston, Barclay's Bank. We were not aware that the Monday was a national holiday to honour the arrival of Haile Selassie, king of Ethiopia. Kingston business was shut down and we were without Jamaican money. However, on Tuesday I visited the main branch of Barclay's Bank in Kingston and requested a meeting with the manager. After waiting for half an hour, I realized the receptionist had not contacted the manger. I questioned her, and she said I could not see him, as I did not have an appointment. I was desperate so I went directly to the manager's office, excused myself, and quickly told him of my dilemma. He told me there was no money in the way of a bank draft in my name. He also advised there had been a post office strike for the last six months. He agreed to contact our bank manager in Calgary, who made arrangements to immediately wire the required money. However, in those days communication was very slow and it would take two to three days to complete the transaction. The Barclay's manager was reluctant to advance me any funds against the impending bank draft. If you think I was desperate previously, I was now in panic mode. I contacted Keith Motto, a car dealer in Kingston whom we had dealt with in 1956. He loaned me fifty Jamaican pounds and we were able to get by for a couple of days.

Jamaicans were an emotional and, it seemed, occasionally irrational people. Haile Selassie was known as king of the black race and many Jamaicans saw him as their hero. When he arrived on Monday and descended the steps of the plane onto the tarmac, there was absolute mayhem. People were everywhere, hanging from balconies, on rooftops, and crowding the fences around the airport. When they saw this tiny black man, they rioted. They burned buildings and fought with the police and wanted to know why, if he were king, he was so small in stature. They accused the authorities of presenting a phony Haile Selassie. After the police regained control, the convoy was commenced to Government House for official greetings.

Nola and I had arranged to have dinner with Joe Kiefer (the fellow who helped me with my driver's licence in 1956) and his wife, Betty. They had moved since I had last visited them and were off Constance Springs Road. We had to pass Government House to get to Constance Springs Road and all roads were blocked off, which required us to wait in a long line of traffic. The temperature that day was 95 degrees Fahrenheit with humidity crowding the same number. Air conditioning was not available in rental vehicles and you normally travelled with the car windows down. However, there were Rastafarians wandering around ours and other vehicles as we sat waiting for the procession to pass. This was a relatively new cult and they appeared most frightening to us, wearing dreadlocks and smoking ganga (marijuana). We rolled up the windows and I suggested to Nola that she look straight ahead and not make eye contact.

When the traffic cleared we continued on our way. We were late for our visit and as usual at 7 P.M. darkness had fallen and I was lost. Nola suggested we stop at a nearby Esso service station where there was a group of fellows dancing around a lit forty-five gallon fuel barrel, smoking ganga, to ask directions. I asked her what she thought might happen to a well-dressed white couple in a rental vehicle? The answer was obvious and we continued on our way. In Kingston the municipal planning leaves a lot to be desired: million-dollar homes adjoined cardboard hovels. I turned up this street and realized quickly that I had made a very poor decision. We immediately turned around and as we were approaching the main street a young black fellow on a motorbike asked if he could help us. He was well dressed and I told him of our dilemma. He led us to the address we desired and refused any compensation. After our evening with the Kiefers, Joe led us back to Constance Spring Road and we returned to our hotel.

We visited with other acquaintances in Kingston whom I had met in 1956, then started for Savannah-la-Mar on the west coast. Although she was apprehensive about her surroundings, once we left Kingston, Nola felt quite safe. She had a feeling that the moccasin telegraph (as we would call it) had gone ahead and said, "Mr. Bob is returning." I contacted George Lynn, who had been the shopkeeper when I worked in Jamaica, and asked him to line up accommodation in the general area. He made arrangements for us to stay at T Water Cottages at Negril Point, owned by Rita Siegried. Rita's husband was the plantation owner in attendance when I logged the first hole in 1956 (remember eenie - meenie - minie - mo?). He had passed away and Rita was, by Jamaican standards, independently wealthy. She had three cottages and served all the local dishes: salt cod, ackee (like eggs), red peas and rice, bread fruit (like potatoes), jerk chicken, wild goat, and much more. On our way to Negril we stopped to visit Williams. He had been our mechanic and had made enough money to start his own shop. He was welding on a Caterpillar track, with his welding mask down, when I tapped him on the shoulder. He raised his welding mask, and said, "Mr. Bob," and the tears started streaming down his face. We hugged and cried. It was a very emotional moment.

After a day or two at Negril, we returned to Sav-la-Mar and went to the golf course, where, fortunately, they were holding their AGM. Everybody crowded around us and there was much hugging and tears. I was the only one of about twenty-five men who had been on the crews to return to Jamaica. We drank and we talked. Joyce and Osmond Hudson asked us to their place for a couple of days. She invited another five or six couples to join us for dinner at their home. Although she contacted the servants of her impending dinner party around 9 P.M., we did not arrive until midnight. Nola and I wondered how the servants could serve such a delicious, hot meal after so many delays. The Hudsons were most gracious hosts and when they found out Nola liked orchids, Joyce made certain there was a garland of them surrounding her plate at each meal.

While in the area we spent a couple of nights at the famous (Shell World of Golf Series) Tryall Golf and Country Club Hotel. It is an absolutely fabulous spot and the golf was fantastic. They had a bar in the middle of the swimming pool.

We visited the Aguilars, the Farquharsons, and many more white families in the area. George Lynn held a party for us and invited all the local white families. They told us later that it was the first time they had ever been in his home. This is partially explained by the fact that he lived common law with a black woman. We were unable to contact Rita at T Water Cottages while we were at the Hudsons, and when we returned, she advised us that Gordon, our gardener in 1956, had arrived and waited for two days. I was sorry we missed him. We toured the remainder of the island, and the evening before our departure, invited the Kiefers to the Blue Mountain Inn for dinner. We departed Jamaica early the next day and flew to Miami.

We were in Miami on Easter Sunday and then flew to New Orleans where we were greeted by Scott and Wilma Childress as well as John and Betty Baxendale. Scott owned Economy Mud Products and supplied Explosives Limited with drilling mud. The Baxendales were good friends of ours from Calgary, John being the chief geophysicist for an oil company office in New Orleans. When Scott became aware of our travel plans, he offered to meet us in New Orleans and show us around. He also asked us to accompany them back to Houston by car and spend a few days at their home.

In 1966 you needed a reservation for all the dining establishments, and Scott knew his way around New Orleans. We stayed at The Inn on Bourbon, situated on the corner of Bourbon and Toulouse streets (the most famous intersection in the French Quarter). We inadvertently stumbled onto a movie set filming a session of "Lassie." The sounds of "cut" brought us to attention and we sheepishly crept away. We had breakfast at the famous Brennans and visited many oyster bars where you mixed your own sauce. Preservation Hall is a very small, smoke-filled, dirty storefront where all the top jazz musicians gather to jam after their regular gigs in bars and eateries. The line starts to form around 7 P.M. and the musicians arrive around 11 P.M. to midnight. What an experience. And we finished the evening by going to Cafe Du Monde (which is open twenty-four hours a day) for strong coffee and beignets. Beignets are deep fried doughnut-like pastry covered with icing sugar. At Cafe Du Monde you rub shoulders with everyone from the tuxedo-clad opera crowd on their way home to the panhandler on the street. With its food and atmosphere, the city of New Orleans is like no other.

From New Orleans we travelled by car with the Childresses to Biloxi, Mississippi, through the swamps to Houston. Scott and I played a couple of games of golf and Wilma took Nola shopping. The Childress children, who were teenagers, called their friends over to the house to hear us talk "Canadian." We flew home to Calgary after a month of happy times and a most fantastic vacation. The only disappointment of the trip was that when we arrived in Calgary on May 15, there were no leaves on the trees. I had expected after our April 15 departure in snow to have a green homecoming.

The "Crazy Man"

During the early sixties the Doodlebug Golf Tournament had a Caribbean theme for the banquet evening. Nola and I dressed up in our Jamaica togs and we headed out. We were invited to the past president's party and I imbibed more than normal. They had an impromptu limbo contest and I, being a past resident of Jamaica, was going to show everybody how low I could go under the bar. Somebody had spilled a drink and my feet slipped. Rather than fall backwards, I pulled myself up and heard two "pops." I had stretched ligaments in both knees.

The beds at the Banff Springs Hotel are quite high and I was hardly able to ease myself into bed. I got up at about 6 A.M. to go to the bathroom and was unable to lift myself off the toilet. Nola came to my aid and helped me up. I was in the golf finals against my buddy, Earle Mahaffy, and I did not wish to default. I contacted John Prendergast, an ex football professional, for his advice on painkillers. He suggested I get in touch with a doctor and get a couple of shots of painkiller. I called the hotel doctor but he refused and told me I could permanently damage my knees.

Nola wheeled me out to the car after retrieving my golf clothes from the locker room and loaded me in the back seat with my legs resting on the seat. When we arrived at the Calgary city limits Nola called our family doctor and he suggested we go directly to the Holy Cross Hospital.

I was admitted and they contacted an orthopedic specialist who recommended tenser bandages and physiotherapy every twenty minutes, twenty-four hours per day. I was in the hospital for over two weeks. The staff had a lot of fun at my expense, telling one and all about this crazy man who tore up his knees doing the limbo. One of the major radio stations picked up the story, so everybody in Calgary knew of my silly injury. The news also found its way to the geophysical community in Australia. I remained at home recuperating for an additional month.

When I returned to work, I found that without the challenge of competition to keep the adrenaline flowing, things became more or less routine in the seismic, construction, and mining industries. XL received a contract in 1965 to supply explosives and ancillary products to Canada Tungsten Mining, located 250 miles northwest of Watson Lake. Although Watson Lake, Yukon, is on the Alaska Highway, Canada Tungsten is situated in the Northwest Territories. I had become very good friends with the mine manager, John Keily, and the assistant mine manager, Mike Babcock.

At that time, federal law only allowed us to haul ten thousand pounds of explosives per truckload. Because of the extreme delivery distance to the mine site, the product price landed was very expensive. Canada Tungsten applied to the federal government to allow forty-thousand-pound truckloads starting at Mile Zero on the Alaska Highway. Both Explosives Limited, the company I worked for, and CIL, the manufacturer, refused to support the request, as it was an inconvenience for them. But because of my personal relationship with the federal chief inspector of explosives and my belief in helping the customer when it was the right thing to do I was able to assist Canada Tungsten in attaining permission to haul increased loads of explosives.

A Setback

In 1966, during my fall trip to the northern branches to prepare for the winter onset of seismic business, I had a frightening experience. I normally left Calgary on Sunday night after supper and overnighted in Edmonton. I had been having difficulty with my neck for some years and the pain in my left arm became unbearable, especially while in bed. I had found that if I tucked my left arm under my back and laid in that position it relieved the pain. The pain this particular night was extreme and I applied my anti-pain position. The next morning I had no feeling in my left arm or hand.

I continued my trip, and when I arrived at Fort St. John, I tried to see a doctor. But with such a large influx of oil workers, they told me it would be two weeks before I could get an appointment. On returning to Calgary some two weeks later, I visited our family doctor. He sent me to a neurosurgeon, Dr. Hepburne. Dr. Hepburne was the son of the doctor who, with a partner, devised the spinal fusion at McGill University in Montreal.

He asked me to push against him, with first the right and then the left arm. I could push him easily with the right arm but could not move him with my left. He asked, "How long have you known you have lost over 75 percent strength in your left arm?" That was news to me.

He told me that I had ruptured discs between the fourth, fifth, and sixth cervical vertebrae and prescribed sessions with a physiotherapist. They placed a harness over my head and under my chin. The harness was attached to a cable and then elevated through a pulley to the floor. They hung weights on the cable to attempt to free the vertebrae, which were pinching the nerves to my left arm. After forty-eight pounds of weights they gave up and I returned to see Dr. Hepburne.

He suggested an operation in which they remove a piece of hipbone and fuse it between the ruptured vertebrae. The alternative was a wizened-up arm with no muscle. I opted for the operation and in November 1966 I had a cervical spinal fusion of the damaged vertebrae. I was supposed to be out of commission for three months but by using the car headrest that I built for myself, I was able to get back and forth to work within three weeks with Nola driving. The odds of recovery in those days were 50 percent but the consequences of not having the operation offered very little choice. The operation was a complete success.

During my spinal fusion convalescence, I started two important personal projects: my autobiography and an investigation into my mother's side of the family, the Burkes.

The Burke Family History

While convalescing in late 1966 and early 1967 I tackled the Burke family history. I recorded my findings in a self-binding book and distributed it to all members of the Burke clan. Nola and I spent three years of my three-week vacations travelling Canada from ocean to ocean and from Dawson City to the American border.

It is purported that the original Burke, George Thew, came to Canada in the early nineteenth century to fight the Americans in the war of 1812. He fought with Brock and was a signatory to the subsequent peace treaty. He was elected to the House of Upper Canada from 1821 to 1829 and founded Richmond, Ontario, which was declared a military town subsequent to the 1812 war.

His great-grandson (my great-grandfather) homesteaded near Okotoks, Alberta, in 1889. From there he and my grandfather homesteaded in the Porcupine Hills, west of Claresholm. My grandfather, Albert Edward Burke, subsequently farmed near Claresholm (where my mother was born) and then moved to Victoria and Larkin, British Columbia. He moved the family back to Alberta in 1918.

While looking into the Burke history and the homesite of my great-grandfather, I came across Christ Church Millarville in the foothills about twenty-five miles south of Calgary. I was looking for a possible Anglican Church that the Burkes might have attended and when I located Christ Church Millarville I felt that it could have been the one. I have not been able to prove my theory.

Nola was raised in the Anglican Church but had agreed to attend the United Church in deference to my upbringing when we moved to Haysboro in 1960. I felt the Christ Church would give her the necessary religious satisfaction and suggested we attend a service.

Christ Church had been built in 1896, using vertical logs in its construction. The vestry at that time was skeptical of this type of construction and refused to pay the contractor for three months. The building withstood the winds and weather, the contractor received his money, and it remains standing to this day.

The priest was Waverly Gant. In addition to being very congenial, he was also extremely wise. Nola and I attended our first service and I advised Waverly that we wanted to become registered parishioners. I mentioned that I travelled a lot and liked to play golf on Saturday and Sunday. As Christ Church was part of a four-church parish, he could only preach at two churches each Sunday due to distances between churches. This meant we attended once every two weeks, leaving two Sundays a month to play golf. Waverly told us: "Your church is where you are. You do not have to visit a church building." He and I hit it off famously. In the early going the members of the church were skeptical of Nola's and my intentions. "What do these city people want from us?" they wondered. However, they soon found out we were there to worship and do whatever we could to help the church survive.

Nola and I heard repeatedly about the "calf program," so we attended the annual general meeting in January 1967. The calf program was a fundraiser for the church and consisted of buying calves in the fall and delivering them to the surrounding ranchers. An extra calf in a herd did not require much extra feed and in the spring the calves were sold. The difference in cost and selling price went to the church and the rancher received a tax receipt for the difference. Somehow I ended up on the vestry, an experience that opened up a wealth of opportunities to assist and participate in the church.

Over the years I became treasurer and rector's warden. The previous treasurer had been in office for twenty-five years. After I had spent a couple of years helping deliver and pick up calves, assisting the treasurer to collect on Sundays, and with Nola helping out at the flower festival and in other areas, the congregation accepted us as members of the church. The treasurer invited us to his ranch for lunch and asked me if I would take over the job. I accepted. This was during an era of reasonably high interest rates. After investigating the church's banking arrangements, I found the bulk of the money was in low-yielding savings accounts and one account did not show up on our financials.

The church always seemed to be in a financial bind. However, with a few changes I was able to invest in higher yielding term deposits. I changed the accounting system and talked Ace's accountants into performing a yearly audit. I had our fixed assets evaluated and presented a profit-and-loss statement as well as a balance sheet. When Nola and I moved from Calgary to Rockyford in 1980, the church was left in fine shape financially. It was merely a matter of bringing the church accounting into the 1970s and I was happy to help out.

As treasurer, I was automatically a member of the Parish Council. I was surprised to find out that Waverly's only pension was a very inadequate sum from the Anglican Church Diocese. I convinced the parish to set up a one-time pension that would allow Waverly, who had been the parish priest for twenty-five years, the ability to retire in relative comfort. As Waverly was living in the parish manse, he would have to secure living accommodations on retirement.

Doctors and religious ministers are sometimes poor administrators of their investments, and Waverly was no exception. I was able to convince him to take advantage of my investment advisors, at no cost to him, and they helped him immensely. I felt a great deal of satisfaction when Waverly and Dorothy were able to purchase a very nice home in Grand Forks, BC. This was an ideal location for Waverly as it allowed him to continue his mountain hiking and cross-country skiing as well as develop a beautiful old English garden. Nola and I visited them yearly until his death in 2001.

While at Christ Church, we had the opportunity to witness

many changes. With additional funds we were able to landscape the grounds. I had retained my survey instruments and used them to survey the property and file a complete survey plot. Nola and I renewed our marriage vows in the church on our twenty-fifth and fiftieth anniversaries. Brenda and John's wedding was held there and their daughter, Jennifer, was christened at Christ Church. We were unable to attend Christ Church once we moved to Rockyford but when we returned to live in Calgary in 1992, we returned to the church in Millarville. They voted me a life member of the vestry and both Nola and I continue to be active in the administration of the church. It has brought us much joy and satisfaction. ■

Part Four

Risk Taker

Action Makes More Fortunes than Caution

X L was always on the lookout for additional business opportunities and in 1967 the company had a chance to procure a line of mining equipment—something I had put a lot of time into setting up. They hired a new fellow to head up this diversification. The problem was that the salary they offered him was about 25 percent higher than Stan's or my salary. Nola was furious: "You make so much money for XL and they pay you peanuts. Why do you not get into business for yourself?"

With Nola's observations still ringing in my ears, Stan and I discussed the situation. We decided to leave XL and start our own explosives distribution business. It was the fall of 1967 and the lucrative winter seismic business was only a couple of months away. It was imperative that we get the business off the ground immediately. Neither of us had investment funds so we had to raise a considerable amount to compete against XL.

We would need to duplicate XL's branch locations and provide equal or superior service.

Stan had some contacts who had made a fair amount of money in the oil industry. I worked night and day to come up with an acceptable business pro forma (I was the business partner, Stan was the salesman). I commandeered my old friend Bill Davis, a senior partner in Price Waterhouse, to help with the pro forma. We met with Stan's contacts for lunch in the dining room of the Palliser Hotel.

It was important that we keep our plans under wraps as long as possible so as not to tip our hand to our prior employer. At this point, surprise was our biggest asset. You can imagine my consternation when I received a call from the hostess of the dining room saying she had found a large file that had been left at our table. Stan's friends had left our pro forma file in the restaurant. They had no intention of backing us.

My leaving Explosives Limited to go into business for myself could not have come at a more inopportune time for our family. My mother and father were living with us due to Dad's poor health. Brenda was attending university and our finances were stretched to the limit. However, Nola had for years encouraged me to get into business for myself and she stood firmly by my side.

I had developed very good business relationships with most of XL's seismic clients and I decided to ask a couple of them to back us to the tune of $225,000, or about $1.5 million in 2009 currency. This was a lot of money, but a necessary outlay if we were to compete on an even playing field with XL. My two targets were Harold Farney and J.C.J. (John) Fuller. I knew Harold better than John but Harold was out of town and time was of the essence, so I presented our proposal to John. Harold was to be back in town in a couple of days. And because I wanted to hedge my bets I also talked to a couple of insurance companies regarding a job in sales. I knew what they wanted to hear and when I completed their questionnaire, the manager of one company called and wanted me to start immediately. He claimed he had never seen a higher score from a first-time applicant. He requested an answer within two days. I called John back and told him I had to have an answer by 5 P.M. on Friday night. Some audacity for a guy with no job and no money! But in the end, they agreed to back us.

Now came the chore of setting up a company with partners. As time was important, we needed equipment, fast. We contacted a small explosives trucking firm by the name of Ace Explosives Limited, who had some explosives magazines and trucks, to see if they would sell their company. They were amenable and we hired Bill Davis as our accountant and Doug Mitchell as our lawyer. Harold and John were currently clients of Bill and Doug. Our benefactors asked us if they could recruit some of their business friends to invest, to lessen their exposure. We agreed, but I insisted that we answered only to them on day-to-day business. They agreed.

When we commenced our negotiations concerning what form the company would take, our backers proposed that both Stan and I should have a financial stake in the company. They suggested twenty thousand dollars each. I borrowed ten thousand dollars from Stan and ten thousand from my dad. We set up the company on a shareholder loan basis so that the company would pay an annual dividend of 8 percent on the value of each shareholder loan. We faithfully paid those dividends each and every year. I knew, regardless of how bad the bottom line was, if we honoured those dividends we would receive no flack from the shareholders. It was critical that we design a formula to eventually acquire the investor shares. It was agreed that we would be allowed to purchase up to 49 percent of the shares at a fixed price per share.

We were off and running, and rented an office above a used auto dealer that happened to be owned by an old school acquaintance of mine, "Skip" Skipstead. It was located on Forty-second Avenue SE, about five blocks east of Macleod Trail. The upper floor had been vacant for some time and required extensive cleaning and rearranging. Our families chipped in and we worked eighteen hours a day to make it habitable. We bought used office furniture and supplies and hired a secretary who had worked for Harold as well as hiring drivers, helpers, magazine keepers, and salespeople. We were in business.

We approached CIL to supply our explosives and ancillary products. Because of the Canadian Federal Competition Act, they were forced to sell and give the same discounts to us as they gave XL. This was $1.50 per one hundred pounds and 5 percent on detonators and other ancillary supplies. Our competitive position filtered down to our prior employer and the "fit hit the shan." They attempted every trick in the book to undermine us but we persevered.

A very important and necessary accessory was a "drive point." This was used with attached "wings" to maintain an explosives charge at the desired depth in a shot hole. The hole was usually full of drilling fluid and if you did not use this accessory, the explosives charge would float to the surface. Since modern seismic exploration had been introduced in Canada, this item had been manufactured using metal. I could never understand how the industry got away with this process. The rules and regulations and common sense did not allow metal other than non-sparking bronze to be near explosives, let alone attached to a charge. A company in Montreal manufactured these cone-like devices with wings. The wings were made out of spring steel and fitted inside the cone and the device was then fitted to the explosives charge. When we attempted to purchase these units from the Montreal manufacturer, both CIL and XL blocked our attempts by threatening the Montreal company that they would switch suppliers.

Without the wing and drive point we were in an inferior competitive position. I knew a fellow in Vancouver who owned a plastic extruding business and between us we worked out a very acceptable product. We patented the item, and then forced the issue of a metal product with the federal chief inspector of explosives. He immediately denied the use of metal in drive points and our competition was now in a pickle. XL attempted to challenge our patent and lost. We sold our product to them, made a profit on the sales, as well as a royalty fee after they started to manufacture their own plastic points. I must admit that this gave me a great deal of satisfaction. I have kept a copy of the original royalty cheque from XL.

Economy Mud Products, a U.S. company, had been supplying drilling mud for the seismic industry in North America for years. We applied for mud supplies from this company but were denied, due to XL pressure on the manufacturer. The Economy Mud products contained peptides, which when mixed with regular bentonite caused mud to expand rapidly to firm up the drill hole walls. We desperately needed to provide a similar product so we contacted a company, oddly enough named Ace Mud Products, who supposedly had a similar type of bentonite. (Bentonite is the basic ingredient for mud.) Their bentonite mine was in Wyoming, where the majority of product was produced for use by the North American oil industry.

Ace Mud's president, his wife, partner, and I flew to Stucco, Wyoming, in their King Air to evaluate the bentonite for use in the seismic industry. Bill Huddleston, the president and owner of Ace Mud, was a real character. He asked me if I had ever attended the International Oil Show in Tulsa, Oklahoma. I had not. My answer surprised him, and as the show was in full swing he suggested that we stop in Tulsa for a few days after our visit to the mine.

I had planned for a two-day trip and was attired in field clothes and my luggage consisted of a shaving kit. First, we flew to Kalispell, Montana, where their U.S. office and his partner's home were located. They had some business to attend to and I outfitted myself with a western suit and shirt. I was wearing western boots. I was ready to take on the Tulsa Oil Show. We flew to Tulsa and had a grand two days. Bill decided to continue on to Lubbock, Texas, where he wanted to visit a friend and customer as well as investigate the purchase of a new plane.

We registered at a motel with adjoining rooms, and I contacted Nola to tell her of my extended plans—rather after the fact. For a lark, I suggested that after I had Nola on the line that they listen in on the conversation. I told them the conversation would track something like this: "Hi Nola. How are you?" Her answer would be: "Where are you and what are you doing for clothes? You realize this little caper screws you for ever owning your own plane." When Nola answered, her conversation was very nearly word for word as I had outlined.

After a test period we ascertained that the Ace mud was inferior to the Economy product. So we turned to Baroid to supply our mud products, with great success. We had not gone to Bariod earlier because their distributor discount was less attractive and we tried to refrain from being involved with large companies. The profit margins were never fantastic, so the problem of our competition lowering prices in an attempt to force us out of the market did not materialize. Within a couple of years, we were very happy with our percentage of the seismic business. Whenever we discussed percentage of business with customers, we always inferred that we were the underdog and if they wanted to retain competition, it would really help if they gave us a little more business.

Starting up a new business was an incredible amount of work, encompassing eighteen to twenty hour days, day in and day out. The secretary we hired had a relapse of tuberculosis and ended up in the hospital. Nola came to our rescue. After working with me for so many years in the seismic business she understood the level of service that was necessary to placate our customers. She was a blessing.

In 1980, CBC's "Business Watch" interviewed me, and Bill Kurchak asked, "What does an entrepreneur do to be successful?" My answer was: "You have to work eighteen to twenty hours a day and mortgage your house, yourself, and your family." Lofty goals demand hard work.

Stan and I split the winter territories. He covered northern Alberta and I covered northern British Columbia. That first winter was rough. The trucks we had inherited from the original Ace purchase and Al Worthing Transport had seen better days.

I spearheaded a convoy to Fort Nelson consisting of a three-ton truck pulling a house trailer, and me in a pickup pulling a

Bob at the inception of Ace Explosives in 1967. It was time to grab the brass ring.

portable magazine. We encountered a terrible blizzard just west of Edmonton and had to shut down for about five hours. We eventually arrived at Fort Nelson and set up our operation. We had hired Shorty Canaday as our magazine operator and we were off. My job was primarily to visit crews in the bush and drum up business. On my maiden voyage I slid down a very icy trail and ploughed into a fuel truck on its way up the hill, immobilizing my pickup. I was very despondent what with the incredibly long hours setting up in Fort Nelson and very little sleep. I was almost ready to say to hell with it and pack it in but it was not in my nature to give up. I had the truck hauled into town and commandeered Shorty along with his pickup and we set off again.

We had a reasonable winter and made a few dollars, but times were not easy. In the summer of 1968, Stan covered Alberta and I covered Saskatchewan. I knew the area due to my previous seismic experience in that province. I purchased a 1967 Cougar, which went like the wind, and I was able to visit a lot of seismic customers in a short period of time. We had a salesman in Saskatchewan, and on one trip I lapped him. Needless to say his term with Ace was cut short. Our summer sales were excellent and we were on our way.

During the spring of 1968 we approached DuPont of Canada to provide our explosives and ancillary blasting products, as it became very clear that our relationship with CIL was tentative at best. Unfortunately, DuPont was not versed in manufacturing seismic product in Canada, so we were exposed to a very long and painful learning curve. Our first delivery consisted of over 250,000 pounds of inferior seismic product. We were fortunate to keep our heads above water during this period. I must say, our customers were very understanding, as they recognized the advantage of having competition in the explosives industry. And DuPont finally provided a product that could compete effectively.

DuPont's credit officer in Montreal was strictly a bean counter. He expected Ace to pay for the 250,000 pounds of inferior product within accepted credit due dates. I reneged and he was about to pull the plug on us. I immediately called DuPont's vice-president of finance and suggested he either remove this fellow from our account or we would have no choice but to discontinue our arrangement with DuPont. The upshot of this brazen act was that the vice-president of finance took over our account directly, and he and I became close friends.

Even with these supply problems our seismic sales were excellent and we kept chipping away at the construction and mining sales, which were much slower in producing returns for us. It helped that DuPont supplied excellent product for these industries.

Seismic provided 50 percent of our gross sales and 75 percent of our profit. Mining and construction provided the remainder of our gross sales but only 25 percent of our profit.

Along the way, I heard from some of our seismic customers that a consulting firm had been questioning them on their explosives suppliers, particularly the DuPont distributor. This was the first indication we had of a survey that DuPont was carrying out to see if they were going to continue to manufacture seismic explosives. I was dumbfounded but decided to play along and see what happened. Eventually, the consultant contacted me and requested an interview. He filled me in on the intent of DuPont's survey and advised that his recommendation was to continue their relationship with us and with the seismic industry. "How do you deal with large organizations such as DuPont when you're a small cog in a big wheel?" I asked him. His answer: "When you feel you have the upper hand, hit them over the head with a baseball bat when they stick their heads out of their hole (figuratively)." Lesson learned! It would come in handy later.

Our first explosives magazine site was located on farm property south of Calgary near the Bow River. It was necessary for us to provide ammonium nitrate fuel-oil (ANFO) based explosives to the construction and mining industry. We did not have a ready supply of this reasonably priced product, and bringing it from eastern Canada was not feasible due to high transportation costs. ANFO is the purest oxygen-balanced explosives available when mixed in the proper proportions. If your mix is not within reasonable tolerances the blast will produce noxious fumes and the rock breaking quality is inferior.

DuPont arranged for us to purchase a truck and mixing unit called a Sudengay Mix truck. It arrived in the summer of 1969 and we were now in the ANFO manufacturing business, but we required a bagging unit. I contacted a couple of old school chums who were the Armco Steel Building agents for Alberta. I found a trailer chassis and we built our bagging unit. The Sudengay mixed the product and moved it through a boom to the roof of the trailer and into the bagging hoppers.

Our first big job for this unit was an Imperial Oil contract for 1 million pounds, for sump blasting in the Arctic. The big rigs required sumps to mix their mud for down hole drilling, and the overlaying ground cover was permafrost. ANFO was an excellent product for blasting in permafrost.

The product was shipped by barge from Hay River down the Mackenzie River to Inuvik. We were able to substantially cut costs by transporting the AN prills by bulk rail car from Calgary to Hay River. We drove the truck and trailer to Hay River and mixed on site. We loaded the bags of ANFO onto rail cars, which were shunted to the barges for the trip down the river. It worked like a charm.

Then, in 1969, we received a call from the U.S. Navy asking us to bid on a simulated nuclear blast at Suffield, the Canadian Forces proving grounds near Medicine Hat, Alberta. They re-

Sudengay Unit at Suffield blast, 1969. Bob designed the bagging unit and the operating plan used in the U.S. Navy's simulated nuclear explosion. The success of the blast made a name for Ace Explosives.

quired a very exacting mixture as explained earlier. Unknown to me all the other explosives manufacturers in North America felt they could not meet these tolerances.

I had visited Cominco's lead-zinc mine in Kimberly, BC, and had been given an underground tour. They were mixing their own ANFO for underground blasting. If the materials were not mixed to very exacting criteria, the fumes could kill their miners. About a mile underground, I was afforded the opportunity to visit the ANFO mix test room. An employee utilized a laboratory scale, a linen handkerchief, a beaker, and a bottle of carbon tetrachloride (CT). His test system was very simple and very accurate. He weighed a given portion of ANFO mix and then added the CT to the mix. The CT separated the fuel oil from the mix and adhered to the AN. The results were filtered through the linen handkerchief, which separated the ammonium nitrate and CT from the fuel oil. The filtered fuel oil was weighed, and, viola, you had the percentage by volume of the fuel oil to ammonium nitrate.

I used this system on the U.S. Navy job and their scientists were dumfounded that such a simple system could work and be so accurate. We built two forty-thousand-pound honeycomb piles of ANFO and a two-hundred-thousand-pound fibreglass enclosure

for two separate blasts. The project was to determine how much damage would be inflicted by these simulated nuclear blasts on structures such as houses complete with foam mannequins representing human beings. Simulated blood bags were inserted in the foam mannequins to visually determine which vital organs would be affected. Of course, the navy had determined how much nuclear energy these two blasts provided. The houses were set off about 1,280 feet from ground zero and two navy scientists were enclosed in a bunker about one hundred feet away from the blast.

Nola and I were invited to the two-hundred-thousand-pound test along with dignitaries from the U.S. and Canadian Governments' scientific and political arenas. We were situated about two miles from ground zero and when the blast was set off you could see the shock wave travelling toward us. It actually compressed

Bob in bagging unit, ANFO plant, for the U.S. Navy simulated nuclear blast, 1969.

the atmosphere. The project was a complete success and Ace had made a name for itself.

Our second explosives magazine site was on the Big Springs Hill road west of Airdrie and north of Highway 1A. We moved to this location in 1971.

Our business steadily improved and we moved our offices to Blackfoot Trail and Fifty-eighth Avenue SE. At this point in time we had branches in Calgary, Spruce Grove, Grand Prairie, Fort Nelson, Inuvik, and Moose Jaw.

Stan, my partner, came up with a scheme whereby Ace would be a compressed air dealer in the north Toronto area for Gardner Denver manufactured air compressors and accessories. The president of Gardner Denver, Canada, was a good friend of Stan's. We rented an office and warehouse just off Highway 410, north of Toronto. We named our endeavour Indair. The year was 1974. Stan and I spent alternate months at Indair attempting to develop a profitable business. Stan's and my methods of organizing a viable business approach were not compatible.

A contract to supply constant air temperature to the large cables stabilizing the CN Tower during construction was a profitable and interesting project. Completed in 1976 it was the tallest free-standing edifice in the world. They were having a problem with the expansion and contraction of the cables that stabilized the structure while it was being built. One of our employees came up with the idea to encase the cables and continuously pump air into the enclosure at a constant temperature, and it worked. It was exciting to be contributing to a structure of that magnitude. On one occasion they allowed me to visit the site and take an open-sided elevator to the top of the tower. Although I am not comfortable with heights over six feet, I persevered and enjoyed the view from nearly fifteen hundred feet above ground.

It was at Indair that Tony Penny joined us and John Saluk (our daughter's future husband) spent a short time as an employee. The volume of business was not as great as projected by Gardner Denver and we closed the door after a year with $250,000 in losses. We had additional losses due to bad accounts but avoided others due to personal relationships we had formed. Business is seldom a bed of roses.

Business Developments in the 1970s

In 1976 Ace had the opportunity to pick up the Canadian distributorship for a four-wheeled, articulated, amphibious bush buggy called a Coot. It appeared to have great potential in seismic and other bush-based industries. Stan and I decided that he would take on the Coot and travel across Canada setting up dealers. This arrangement did not work out, and Stan and I agreed that he would move to the west coast and set up an independent explosives distribution company.

One amusing incident came out of the Coot situation. Stan had set up an ex NHL player as distributor on Vancouver Island. He was provided with a few Coots and parts, and that is the last we heard of him. He refused to pay for the product or parts and his account was over six months old. I caught word that he was coming to a summer hockey school in Calgary. I went to the rink where he was instructing and advised him that if he did not pay up NOW, with a certified cheque, I was going to expose him in the press. We received our money.

However, the Coot distributorship allowed me to travel to the U.S. for meetings with the manufacturer. Originally, they were manufactured in Dallas, Texas, and subsequently, in San Francisco. I finally achieved my goal of visiting New York, New Orleans, and San Francisco. The CEO of the company was an exceedingly nice man and insisted I bring Nola on my first visit to their factory. He and his wife toured us through the San Francisco area, including Chico, north of Oakland, where we visited and joined very dear friends of theirs for dinner. These folks ran an olive ranch and sent us home with a couple of jars of their choice olives. Nola and I visited Fisherman's Wharf, Ghirardelli chocolate factory (started in 1852, the oldest chocolate factory in the U.S.), and rode the famous streetcars. We visited Sausalito, just across the Golden Gate Bridge, and Napa Valley, as well as the abandoned prison on a rock island in the bay called Alcatraz. Over the years, Nola and I have visited San Francisco and area many times, enjoying excellent restaurants in North Beach and Chinatown.

I hired Al Weeks from DuPont, as Ace's vice-president. Al was western Canadian manager for DuPont's explosives operation. When Stan was apprised of this development, he requested information on his position within Ace. I offered to purchase his shares. After some negotiations I purchased Stan's shares and made a donation to his campaign as a member of parliament in the Kootenay riding. He won and entered Joe Clarke's Conservative Government in 1979. He was defeated in 1980, re-elected in 1984 to Brian Mulrony's Conservative Government, and was defeated in 1988. After purchasing Stan's shares I exercised my share purchase option and eventually owned 49 percent of Ace.

I had made it a "must" to be as comfortable as possible with our banking arrangements. I reasoned that as I was the customer, the banks should, within reason, acquiesce to our position. Of course, as we all know, banks do not operate in that fashion. However, I was determined not to let the banks browbeat me and I carried out my negotiations with them on my terms. Business associates told me that with my attitude we would find ourselves without adequate banking arrangements. This was not the case.

Our backers were affiliated with a Calgary branch of the Toronto Dominion Bank at Eighth Avenue and Eighth Street. The manager was a rather pompous individual who resembled Errol Flynn in looks and manners. After a few months of business were under our belt, and, quite frankly, not going very well, he announced to one of our minor backers, that we, in his opinion, were going to fail. Not a word from the manager to us. I immediately contacted our major backers and advised them that we were changing banks. Fortunately, they understood our position. My partner suggested I contact the Canadian Imperial Bank of Commerce (CIBC) where his experience had been satisfactory. We continued our banking arrangements with them until I purchased Stan's shares.

CIBC had appointed a senior employee in the main branch to supervise our account whose name was Trudeau. His name alone was a strike against him. On occasion he commented on my open-necked shirt. I wore a tie when calling on head office customers and major suppliers. Mr. Trudeau did not warrant a tie. In my opinion, he was very narrow minded, and with the plans I had in mind for Ace, I knew he would not be supportive. After procuring other banking arrangements, I let Trudeau know that we were moving our account to another firm.

CIBC's senior vice-president for southern Alberta contacted me, requesting a meeting. I suggested that if he could see me that evening at 5:50 P.M. I would be pleased to talk to him. After the initial introductions he expressed his surprise regarding our actions to leave his bank. I pointed out to him that at any given time our company was running some 3 million dollars through his bank, yet this was the first time he had asked to meet with me. I also commented on Trudeau's habit of "talking down" to me. His request for reconciliation, I told him, was too little, too late.

Nola and I had dealt with the Bank of Montreal for years. The young manager in our Haysboro branch told us that his boss was anxious to handle our company account. After several meetings we cut a deal that was far superior to our previous arrangements with CIBC. Our arrangement with the Bank of Montreal ran very smoothly and it was this bank that approved the additional funds to purchase land and build facilities near Rockyford, which I discuss below.

I was invited to many bank functions and met all the top management staff including the Alberta vice-president, Simon Kowenhoven. He was a true gentleman and appeared to understand our position. The banks had a habit of lending adequate funds for startup or expansion but if you asked for a 1-million-dollar operating fund, they would attempt to cut it to seven hundred thousand. I would not accept that position because if you were unfortunate enough to come up short a couple of months they would call your loans.

Sometime after we opened our Rockyford facilities, the Bank of Montreal business manager whom I was dealing with left the company for a better position. The new business manager was worse than my old "friend" Trudeau at CIBC, so I contacted Kowenhoven and informed him I could no longer support the new business manager. He told me that because of internal Bank of Montreal problems, his hands were tied. Kowenhoven contacted me about a year later and asked me to sit on the Bank of Montreal's Inaugural Small Business Advisory Board for Western Canada. I was the only non-customer board member.

When we relocated in 1980 to the Rockyford area, the only bank in town was CIBC. It did not take a genius to figure out our next move. Fortunately, the bank manger in Rockyford had been wooing us for some time. Our arrangement with CIBC was even better than that with the Bank of Montreal and to this day Nola and I are CIBC customers.

CIBC's oil and gas manager requested that our account answer to his division once we moved to Rockyford. Upper management declined his request and placed us within their general business department. This did not affect us until they assigned a credit manager to our account who attempted to play the old "cut you down on operating funds" game. I tried to reason with him to no avail and I finally contacted Dave Lowry, VP for Alberta. After some discussion I told him that it was either a different credit manager or we were moving our account. Lowry agreed to appoint a different credit manager and we continued to bank with CIBC. Who says you cannot shop banks?

With Al Weeks's experience, a concerted effort by all employees, and sticking to our basic business of providing service to the seismic, mining, and road construction industries our business improved dramatically. We were up to twenty-five employees and had changed to contract truckers, hauling from DuPont's North Bay, Ontario, explosives plant in forty-thousand-pound loads to Calgary and points north and west. Some of our branches delivered explosives using a combination of Ace owned and contract trucks.

DuPont had, over the years, fine-tuned the manufacture of seismic nitroglycerin explosives and had introduced, at our request, a water gel based explosive for the seismic business. This product had been developed for the mining and construction industry with excellent results. We named the water gel product Energel. Energel was a much safer product than nitroglycerin-based explosives, which could detonate when exposed to extraneous conditions such as fire, accidental concussion, or electrical storms.

In the early 1970s, DuPont's North Bay explosives plant blew up, and the company was unable to manufacture nitroglycerin-based explosives. They promptly arranged with CIL to enter a tolling arrangement—a method whereby CIL manufactured and boxed explosives to DuPont's specifications—which allowed a continuing flow of products.

Ace had previously been able to supply Gulf Oil with a special product for sump blasting. Sump blasting is the method of excavating in permafrost for water storage at Big Rig locations. CIL originally refused to manufacture the special product, three-inch by ten-pound cartridges. It was a very lucrative business for Ace and DuPont. However, subsequent to the DuPont plant explosion, CIL refused to "toll" the product for us and in fact started to manufacture their own three-inch by ten-pound to capture Gulf's business.

DuPont informed us that they had no control over the situation. The business was huge and in the past we supplied up to four-hundred-thousand pounds of product per year.

I was not about to give up this business without a fight. I contacted Hercules Powder Company in Carthage, Missouri, and they agreed to manufacture the required product. I secured the Gulf business and then travelled to Carthage to oversee the manufacturing of the product. Apparently, Hercules had very little experience in manufacturing nitroglycerin-based explosives for use in the far north. Temperatures in the Arctic can bottom out at minus 80 degrees Fahrenheit, at which time the nitroglycerin separates from the other binding ingredients. Therefore, there are small globules of pure nitro that might cause an explosion when inserting the brass punch to accommodate a detonator.

Fortunately, I had acquired the exact percentage of methyl glycol needed to deal with this dangerous situation. To make certain that the proper formula was employed, I remained in Carthage until the four, one-hundred-thousand-pound rail cars left the plant. Only two days after my return to Calgary, I received an urgent message to meet with DuPont officials in Toronto. Apparently, CIL had "spies" at Canadian Customs who informed DuPont of our importation of explosives from Hercules. DuPont was very upset and claimed this type of action was not condoned. As our phone conversation did not fully answer their questions, they asked me to come to Toronto. When I was ushered into the boardroom, I was surrounded by DuPont personnel from the president to the manager of explosives to numerous underlings. I sat at the end of the table opposite the president. After my explanation of losing a major customer, which affected our profitability, they did not back away from their opinion that I could not continue to purchase explosives products from other manufacturers. They threatened to void our distributor agreement, and I suggested it was not worth the paper it was written on.

It became evident that we had reached an impasse in our discussions, so I folded up my file and thrust it into my briefcase. As
I rose from the chair I let them know that Ace would be distributing explosives in western Canada and the Territories whether it
was in company with them or some other explosives manufacturer. I relied on the lesson I had learned from their consultant a
few years ago. My adrenaline was pumping as I realized I had put
our company's future on the line. But in the end, we were able to
come to a pleasant conclusion to this deal and agree to financial
conditions. We were paid by the customer in fifteen days and were
not required to pay Hercules for ninety days. This allowed us to
place the funds in a term deposit for seventy-five days when the
Canadian dollar was at a 10 percent premium to the U.S. dollar.
We made money at every turn.

Over the years, Ace Explosives Limited was the forerunner of
many new and imaginative products and service ideas that helped
us become very successful. We introduced profit sharing for all
our employees and distributed 15 percent of the net profits before
tax and my cut. We also provided all employee benefits, up to and
including employee contributions to the Canada Pension Plan.
Our company was very family oriented and employee loyalty was
supreme. We moved our offices to larger premises on Glenmore
and Blackfoot Trail in 1972 and moved once again in 1975 to a
location off of Blackfoot Trail and Seventy-seventh Avenue.

During this era, we increased our seismic business to 50 percent and our mining and construction business to about 25 percent of the mining and construction sales in our market area. As
well as our head office location in Calgary we established
branches in Spruce Grove, Valleyview, Grande Prairie, and Edson
in Alberta; Dawson Creek, Fort Nelson, and Revelstoke in British
Columbia; Hay River, Fort Simpson, and Inuvik in the Northwest
Territories; Whitehorse in the Yukon Territories; and Regina and
Moose Jaw in Saskatchewan.

The seismic industry was the first to be affected by the vagaries of government and other influences on the oil industry as a whole. On a downturn we would hunker down and cut our staff as required. All employees were aware that the company would not stand by in bad times and lose money. If conditions dictated a reduction in income, then there was the possibility of staff reductions. They were also aware that empire building within Ace was not acceptable. I had been exposed to companies where the success of the organization was the number of employees rather than the bottom line, the net profit.

However, we did pay the best wages, provided superior employee benefits, and included a very attractive profit-sharing plan. I found cutting employees, to control our expenses, very gut wrenching, but I always gave them six months to a year severance pay. In every organization, you have employees who are marginal. All of our employees were not top grade. The employees who had marginal results on their most recent yearly reviews were the first to be selected for release. We sold all but a couple of our company-owned trucks and hired contract trucks. During downturns, management as well as other staff were pressed into service, delivering product and service to our customers as required.

Union Talks

In the early 1970s DuPont succeeded in obtaining a contract from Kaiser Resources to establish a bulk ANFO and water gel plant near Sparwood, British Columbia. This was a very large contract and the first major bulk plant for DuPont in western Canada. The amounts of explosives and accessories used were huge.

Ace was offered the opportunity to supply limited amounts of accessory items, which were very small potatoes. However, DuPont was unable to stop the United Mine Workers (UMW) from organizing their operation, even though the bulk plant only operated with one foreman and six other employees. Rather than unions, all DuPont's North American Operations operated with company-organized "Employee Interest Groups" that supposedly looked after the interests of the employees.

The UMW did not back down, and subsequently the DuPont employees voted for the union to represent them in future labour negotiations. DuPont was between a rock and a hard place, as recognition of the UMW would influence all their business, not only explosives but also chemicals and other products in North America. In other words, the tail wagging the dog. They could not accept the idea of half a dozen employees pushing their entire organization into the unionized world.

When it became apparent that DuPont was going to lose the battle with the UMW and before the union successfully won representation for the employees, DuPont approached Ace and asked if we would take over the employees. They requested that I personally carry on negotiations with the UMW. They would pay us a handsome override and allow us to supply all their accessories at a reasonable profit. We accepted.

Negotiations were rather hilarious. The mine was located at Sparwood and DuPont would closet their human resources crew thirty miles away in Fernie. I would drive back and forth each evening after negotiations had concluded to plan our line of attack for the next day. It did not hurt my position when the UMW representatives realized that my dad had been a top union man in Canada and one of the originators of the Canadian Union of Public Employees. For some odd reason, I felt a slight kinship.

The initial and subsequent negotiations usually spanned a week and sometimes were very physically and mentally taxing. I did not always agree with DuPont's line of reasoning, but in the end I had to represent their point of view.

This continued on for six years and proved to be a very profitable arrangement for Ace. I enjoyed the entire exercise and learned a great deal about dealing with people who would only back down if they were pushed to the wall. We never had a strike.

The Move to Rockyford

I travelled extensively in the United States and had observed that many explosives distributors had established their offices and explosives magazines on purchased farmland. These locations afforded exclusion from populated areas and were adjacent to transportation corridors. This allowed explosives delivery trucks to circumvent urban areas. With the information garnered in the U.S., I started a search for a similar location for our Calgary office and explosives storage. Preferably, the site should be at least fifty miles beyond the city limits with adjacent transportation corridors.

After a few weeks of driving back roads near Calgary, I commandeered a good friend, Bill Gillot, to use his plane. I spotted the ideal location about ten miles northwest of the town of Rockyford, Alberta, just off Highway 21. The site would allow us to travel east, north, south, and west without having to travel through Calgary. On investigation I found there was three quarters of a section available for five hundred dollars per acre. I figured land cost, eight underground explosives magazines, a manufacturing site, an office, a home for the farm and magazine manager, a home for Nola and me, as well as a mile of roads built to county standards would cost around 1 million dollars. That was in 1979, so costs in 2009 would probably be around 3 million dollars.

I phoned my bank manager on a pay phone from the bar in Rockyford and he said, "Go for it." I talked to my backers and suggested that I purchase the remaining 51 percent of Ace. I felt the Rockyford relocation was not a practical investment for them at their age. They agreed and I proceeded with my plans, as the location was ideal. It was situated about three miles south of the Rosebud River where the coulees start their descent to the river. Explosives energy, if detonated above surface, expands out and up. The Federal Government Mines and Mineral Resources Branch governs all explosives storage locations, with a chief inspector of explosives exercising the regulations. The site met all the necessary elevation and horizontal topography and distances. When completed it was said to be the most up to date and safest explosives magazine site in North America.

We started construction in April 1980 and the site was completed by November of that year. There was a surge in the general building industry, which caused no end of grief, but we fast-tracked our plans from inception to completion. This was made possible by hiring a construction manager. Ace had purchased two housing units with two apartments each in Strathmore so that our employees would have some place to live temporarily until housing could be constructed in Rockyford or Strathmore. The fellow who built these units and sold them to us was Carl Eisner, so we hired him as construction manager. He had a handle on immediate availability of the necessary trade personnel and was an excellent draftsman.

Carl's expertise in drafting helped us immensely, as our architects could not keep up to our construction schedule. Carl would work all day at the site, then go home and draft up the next day's work. I gave him a thirty-five-thousand-dollar bonus on completion of the project. Tony Penny, a jack-of-all-trades, and his family lived in the lower unit and Nola and I lived in the upper unit of one of our Strathmore housing units. Tony and I spent eighteen hours a day, seven days a week supervising construction at the site. Tony was a practical guy who had travelled all over the world working on sugar plantations in Jamaica and sailing the oceans. He had a mechanical mind.

We constructed the explosives magazines by excavating into a hill in a semicircle. We used Armco half-circle galvanized steel, mounted on concrete pads, as the roof and side walls and the front and back walls were of concrete construction. We then back-filled dirt over the entire structure. The semicircle location allowed us to maintain our regulated, safe distance between the magazines and the void between was filled with compacted soil. They were located in a circular coulee with surrounding hills in all directions.

Bob at Ace Explosives, Rockyford magazine site, 1981. The magazine site was one of the safest in North America.

If there were an accidental detonation, the energy would be forced in an upward direction.

My experience as a surveyor allowed us to economically survey the property, buildings, and roads. We decided to clad the exterior of the office and houses with cedar and rock. In the spring of 1980, Tony and I travelled to British Columbia and contracted all our building materials direct from the mills. We had a magazine site at Revelstoke, located in a quarry, that mined mica schist, and it provided our stone. We back-hauled building material from BC on explosives trucks after delivering their loads to mines and construction projects in the area. All our buildings were constructed of the best fir, cedar, plywood, shakes, and mica schist.

Home, Sweet Solar Home

Nola and I built a passive solar home on the site. It was located on a quarter section of the purchased land. I had read about a passive solar home in a *Popular Science* magazine that included pictures and schematics. The basic plan was to build a 2" x 6" outer shell insulated to R18 factor and an inner wall 2" x 4" to R12 factor leaving a one-foot plenum between the walls. The roof was constructed to an R36 factor. Only one small window, located in our bedroom, was exposed to outside elements.

The size of the house was determined by our billiard table. We had acquired a full 6' x 12' snooker table, which we set up in our basement in Haysboro. It was so large that we had to use cut-off cues in certain tight spots next to walls. A 6' x 12' table requires six feet of free space in all directions. We built a full-sized billiard room, with a turn of the century office, complete with a simulated barbershop in one end (this combination of pool hall and barber shop was common in the small towns of western Canada up to the 1950s). For the ceiling, we cleaned and repainted rejected tin ceiling tiles acquired from Heritage Park in Calgary. We wainscoted and pasted flowered wallpaper on the walls. We added a Station Agent No. 8 coal-burning stove (the type they used in western Canadian railroad stations across the prairies). I acquired this unit many years ago at a Lethbridge metal scrap dump for three dollars. Using a wire brush to remove the rust, we repainted it with stove black. The office was furnished with a lawyer's leather horsehair couch, wall crank telephone, roll top desk and chair, and multiple-layer windowed bookcase along with a file cabinet, all in oak. It was an impressive sight.

Off the billiard room was our wine and vegetable cellar, which was excavated under the courtyard. It contained storage for four hundred bottles of wine and proper bins for vegetables and canned goods. The remainder of the lower floor included a fully equipped craft room for Nola and two bedrooms with an adjoining bathroom. The billiard room, craft room, and one bedroom were complete with sliding patio doors that exited to the solarium.

The main level consisted of a raised living room with a formal dining room, entertainment area, master bedroom with en-suite bathroom complete with a steam shower with a half bath as well as a kitchen, storage areas, pantries, and a laundry room with a deep freeze. Off the kitchen was a deck and spiral stairs to the lower floor of the atrium. On the south wall of the master bedroom we installed two large windows (acquired years ago when the Holy Cross hospital was being renovated and the chapel closed). A passage door off the bedroom led to a deck in the atrium.

The outside measurements were 56' x 40' and there was a 56' x 12' atrium on the south side. The home was two storeys, with the first storey built into a hill so that there was a "walkout" through the atrium. The atrium was twenty-five feet high and had two feet of sand under a floor of concrete "paver" bricks like those used to construct sidewalks. Passive solar homes generally use large barrels of water to hold solar heat gained during the day to be generated throughout the house during the evening. I recalled that when I was a child playing at the beach the sand during the day could be extremely hot to the touch while at night, when you dug into the sand, the heat had been retained. So sand it was. The atrium had a total of twenty-four 4' x 8' sealed window units to attract the solar energy. The home was placed at the maximum angle to attract the most sunlight on the shortest day of the year.

During the daylight hours, the sand and paver bricks absorbed the solar energy and at night distributed the heat through the plenum around the inner house. As a matter of interest this was known as an "envelope home" and the designers claimed it was "a time tunnel to eternal spring." In fact, it was. The outside temperature could be minus 30 degrees Fahrenheit and on a sunny winter day the temperature in the solarium would be plus 70 degrees. Often, in the winter, with the sun reflecting off the snow-covered ground, I would lounge in the solarium with a good book, clothed in a swimsuit. We installed three large exhaust fans in the apex of the shanty roof, which during the summer picked up cool air off the north wall, pulled it through the house, and expelled it to the exterior.

Of course that many windows (this before the time of the energy efficient windows we currently enjoy) lost a lot of heat during the winter evenings and on cloudy days. We installed two electric wall furnaces in the solarium and set them at 50 degrees Fahrenheit. We investigated insulated drapes for the windows but the cost worked out to a twenty-five-year payback. Our main heating source was a fresh air fireplace that could burn coal. All inner rooms were equipped with a base electric heater as backup if required. The base heaters were seldom used.

The house was built so that the roofline on the north wall was about eight feet from the ground and rose to twenty-five feet on the south (as the south was a walkout the roof line was only two to one). We provided a courtyard between the house and a three-vehicle attached garage to the north, which also sported a shanty roof. Therefore, the north wall of the home was not subjected to the prevailing north winds of winter. The garage and the house were separated by an air lock. On entering the air lock, you closed the outer door before you opened the door to the inner house, therefore no cool air entered directly into the house.

It was an extremely pleasant house to live in and we felt very fortunate that we did not require natural gas for heating. We had many visitors who were curious about our unusual home. We investigated wind power, but Calgary Power would not purchase our excess energy.

Nola was responsible for the landscaping and, along with the landscape architect, worked extremely long hours to complete the task. She was also responsible for the interior designs of both houses and the office.

Upon completion of the "farm" and business buildings, I contacted Omer Patrick, a school and hockey compatriot who was now owner and president of Atlas Mines in the Drumheller Valley. I prevailed upon him to supply us with coal for our fresh air fireplace. They recently had shut down the mine but he said he would contact the person in charge at the mine and if we were prepared to bring the coal up from the level where it was stored underground, we could have all we wanted. Tony and I took the farm grain truck and loaded it with coal. The original farmhouse on the property was quite dilapidated but the floor and basement were sound so we stored the coal in the basement excavation.

We moved into our solar home in early November. Brenda and her husband, John, arrived from Sherwood Park to visit us in our new digs. The weather was fantastic until Sunday afternoon when they were ready to leave. A blizzard blew in, and they were ill equipped to travel in such inclement conditions. They were clothed in shorts, T-shirts, and light shoes. Nola sent me to the garage to fetch a couple of sleeping bags and warm clothes for their journey home.

As I explained earlier, the garage shanty roof increased in

The Rintouls built a passive solar home at Fieldstone Farms. They had many curious visitors anxious to see a home described by one journalist as "a time tunnel to eternal spring."

slope from north to south allowing a storage mezzanine where our sleeping bags and other equipment were located. The construction crew had left a twelve-foot ladder with a wonky leg. I hoisted the ladder and leaned it against the mezzanine floor. I proceeded to climb the ladder and was just about to step off the top rung when the ladder gave way. I fell ten feet and my head hit the concrete floor, flinging my glasses about fifteen feet away. My thoughts as I was falling went something like this, "Oh my God, I finally have my dream home and location and I'm about to kill myself." Fortunately, I was wearing a thick down-filled jacket and as I landed on my back with arms and legs outstretched, the sum total of my injuries was being unable to breathe for a moment and a slight concussion. I crawled over, replaced my glasses, and continued to crawl to the air lock and to the bottom step of the landing. I had left the door open to the inner house. I was really sucking for air and called out for somebody to help me. Nola called back, "Oh, Bob. Quit clowning around." Then she came to the top of the stairs and saw my dilemma. Knowing how to fall and the thick jacket had prevented more serious injuries.

My first thought was for my cervical 1966 spinal fusion and I wondered if I had damaged my spine. I visited the doctor on Monday and after many tests he declared me fit to carry on. About a month later while coaching the bantam hockey team, I was attempting to teach the boys how to defend the front of the net. I was skating backwards and my skate blade caught a piece of bare concrete and I fell again, hitting my head. This time my arms went numb and I was really frightened. The numbness quickly subsided. The next day, the doctor laughingly asked, "Do you have a death wish?" Again, after many tests, I was declared fit. I must have a thick skull. Some say they knew that all along.

Ken Thompson, a dear friend and customer, witched our well sites at Rockyford. He picked the locations and when the driller arrived, he said, "There's no way you're going to get water at those sites." The location at the office and our home hit water at ninety feet. The wells produced multiple gallons of water per hour. We had tapped an underground stream. During our house construction, the workers left for the July long weekend, forgetting to turn off the pump. When we returned on Tuesday morning it was still pumping water, full bore. We installed a three-hundred-gallon water storage tank in the lower level that included a bladder. If the power ceased, the bladder would supply enough pressure to look after our water needs.

In 1986 southern Alberta endured a vicious May snowstorm that toppled large metal power standards like matchsticks. We were without power for ten days but because of our solar heating and special water storage, we survived very well. We operated a small power plant, which maintained the fridges and deep freezes and we had a large kerosene-fuelled camp stove in the garage for cooking.

Our solar home was a delight and I was very proud of having engineered the entire concept and of Nola's design and landscaping skills.

Good Times in Rockyford

Once we settled into our Rockyford location there was no holding us back. Our profit margins improved dramatically and the explosives distribution business flourished. Our accountants worked with us to develop a profit-sharing plan that would satisfy the federal tax office. The plan allowed us to pay half of the profit share directly to the employee and half into a federal government Deferred Profit Sharing Plan. The latter portion was a big tax break for the company and the employee.

We agreed to pay 15 percent of net profits before taxes and dividends to me. The employee could elect to place his or her entire profit share in the deferred part of the plan. Our office clerks were earning up to four thousand dollars in a good year, in addition to

Fieldstone Farms gate entrance, 1985. The unique farm/ranch designed and built by Bob and Nola became their dream home.

their very generous salaries. As previously noted, we provided all their benefits including life, health, dental, and insurance. I expected 100 percent loyalty and am happy to say that I got it. Nola wondered what would happen "if an employee was ungrateful." My reply was, "That's life. You remove the employee and carry on." I remember one instance when I was discussing a special project and wanted an employee to investigate. He responded, "What do I get out of it?" "You get to keep your job," I told him.

We lived in our home at "Fieldstone" (we named the farm after my great-great-great-grandfather Burke's home near Richmond, Ontario) for about twelve years. We were active in the Rockyford Community and contributed in many ways. We were a sponsor at their yearly rodeo and on the twenty-fifth anniversary donated a hand-tooled saddle for the winner of the saddle bronc competition.

That was the same year that a couple of our employees (local farm hands) talked me into joining them in the wild horse race. I was tie down hand, another was ear biter, and the third was the rider. In a wild horse race they let out enough unbroken horses into the arena to accommodate as many three-member teams as are competing. The tie down man lassoes a horse and holds the animal's head down. The ear biter then bites the downside ear, which is supposed to tame the horse, and, at this point, the rider mounts the animal bareback and steers the horse to the finish line. Obviously, the first team over the finish line wins. Well, I was successful in bringing the head down and the ear biter did his job but just as the rider was ready to mount, the horse brought his head up and clipped the rider under the chin knocking him senseless. Game over.

One year the rodeo committee asked me if I would be the target in a charity dunk tank. For the uninitiated, the target sits on a flat piece of metal that is attached by levers to a metal plate a little larger than a baseball. Contestants purchase three balls for five dollars and fling the balls individually at the plate. When a ball hits the plate the force actuates the levers and dumps the target into a large tank of water. I was a popular target, especially to our employees and others who enjoyed seeing a local businessman get soaked.

Bob and Nola were active in the community of Rockyford. Here Bob volunteers to take a dunk for charity at the Rockyford Rodeo, 1982.

Each year there was a parade on the Saturday morning of the rodeo to kick things off after a free pancake, bacon, sausage, and egg breakfast. The parade was very short and they came down Main Street twice. Ace owned a fire engine, which we entered into the parade. It had been in service at the Whitehorse airport in Yukon during the building of the Alaska Highway. Our staff rode on the fire engine and tossed candy to the kids on the parade route.

I coached a Bantam hockey team in Rockyford for a couple of years and was active around the arena contributing time and money for improvements.

Nola and I contributed a trophy to the 4H club for public speaking and were part of the committee of judges. Rockyford did not have a seniors hockey club so Ace supplied sweaters and socks for the team: the Rockyford Aces. I played until I was fifty-five, at which time I had to choose between continuing to play hockey during the winter or going to Arizona where we had bought a house in Sun City. I'm no fool and the warm weather helped my arthritis.

Ace Explosives, Sold!

During this period, I worked on plans to purchase the remaining shares of the company from the original investors. Our accountant, Bill Davis, suggested that we hire an independent accounting firm to determine the price per share. It is very difficult to arrive at a value per share for a private company. After a couple of weeks of delving into our records, they suggested a price of between $79 and $109 per share. As our investors had been so good to us, we offered and they accepted $109 per share. After another trip to the bank, this transaction was completed. We gave each of the partners a silver bowl, suitably engraved, thanking them for their faith in us and for allowing us to grab the brass ring.

Without the fantastic cooperation of my wife, Nola, and her absolute dedication to me and to Ace, I would still be working in the field for a seismic contractor. It was very clear to our employees that Nola was an equal partner in all our endeavours.

After years of providing employee profit sharing, in cooperation with our accountants, and the grace of the Canadian Revenue Department, we developed a proposed employee purchase program. The company would continue the profit sharing, as well as contribute one-third of the funds required to finance the employee purchase. The employees would contribute the remaining two-thirds on a fixed dollar formula. After ten years the employees would own the company.

Nola and I departed for a vacation, and I asked Al Weeks, our vice-president, to visit our branches and discuss the proposition with our employees. I suggested that they might feel more comfortable asking delicate questions relating to finances and other matters of our accountants and lawyers, without my presence. However, I insisted on an answer one way or the other on my return.

When I got back from my vacation Al asked me if the employees could buy me out. This thought had never entered my mind, and I asked for some time to discuss this proposal with Nola, our accountants, and our lawyers. It was determined after consultation with our professional advisors that, yes, we would sell at a definite price, which was non-negotiable, and all monies must be in our hands at closing. Also, I would be retained as a consultant for one year. These terms were accepted.

When the twenty-five employees approached the bank, it insisted on Nola and I carrying the cost of the land and loaning the new owners $250,000 for two years. That was acceptable to us, as the current interest rate at that time was around 15 percent. The new owners completed their obligations within one year. They retained the 320 acres of land, which included the office, explosives storage, manufacturing facilities, manager's house, storage barn, and roads.

Nola and I retained two hundred acres and our home. We were able to negotiate a term deposit with CIBC at 15 percent for five years. The selling price was substantial and the term deposit was very advantageous. The purchase was beneficial for the employees as well. After a few years their manufacturer bought them out and they made a substantial profit on their original investment.

Ace Explosives no longer exists. After a number of buyouts, Dyno Explosives, from Europe, purchased and amalgamated a number of explosives companies in North America. Ace was one of those companies. I was often asked why I had not included my family in the company. Quite frankly, no members of my family were interested.

Over the years we have attended a number of "get-togethers" with ex Ace employees and in every instance they talk about the feeling of working within a family. With the profit sharing and benefits that our employees enjoyed, there was never any "empire building." We operated with a maximum of twenty-five employees spread over western and northern Canada and just before we sold out in 1984, Ace was "writing" nearly 12 million dollars in business.

One of our big coups had been the supply of explosives and accessories to the uranium mining companies in northern Saskatchewan. We commanded 100 percent of the business. The largest company was Gulf Minerals. Its hierarchy consisted of John Keily and Mike Babcock (originally managers at Canada Tungsten). Keily eventually became president and CEO, and Babcock moved up to general manager.

Gulf hired a large engineering organization to construct its plant at Wollaston Lake, and when Keily arrived the inventory and records were in terrible shape. We were in the throws of a petroleum exploration downturn and were going to lay off a number of employees. Keily was aware of our situation and contacted me to give Gulf a hand in straightening out their inventory problems. He asked me to provide up to ten of our employees, to give them a hand, not to skimp on wages, and to include a generous administration fee. When Keily was at Canada Tungsten, I had supported the company in its quest for larger explosives truck loads. It goes to show you "what goes around comes around."

I instituted a manual for Gulf Minerals's deliveries, as the distances were extremely long and there were shipping problems. Because the bush road north of Prince Albert was narrow and dangerous, Gulf contracted a trucking company to transfer all their supplies from that point to the mine site. They were equipped with radio contact between the trucks and the mine. We transported explosive supplies to a drop off point where Gulf's contract trucks hooked onto our loaded trailer and we hooked onto the empty explosives regulated trailer. The contractor continued on to the mine and we returned to Rockyford with the empty trailer. Due to extreme distances and cold weather timing was everything and responsibilities were shared. The manual carried the day.

Eventually Gulf, DuPont, and Ace developed an on-site bulk plant. We received an override on the bulk plant production and all sales of explosives accessory items. All in all, it was a very satisfactory and profitable contract. Other uranium mines in the area were much smaller than Gulf and did not require the minute planning.

Building and operating a business brought me frustration, challenge, and, in the end, immense satisfaction. All I can say to those of you who want to grab the brass ring is be prepared to work up to twenty hours a day, commit completely, and make absolutely certain that you have an equally committed spouse or significant other. Also be a person of integrity and honesty with a strong belief in your personal principles. You also must be generous with your employees and not ask them to carry out tasks that you are not prepared to perform yourself.

Nola and I are very proud of the fact that we started off with absolutely no funds and have achieved peace and independent wealth. We also have the blessing of good health and a wonderful family. ■

Part Five

Giving Back

Nola and I have worked hard for what we have attained but I, personally, more than Nola, have also played hard. If necessary, I would start the day at 5 A.M. so that I could take a customer or friend for a game of golf in the afternoon. I always took my clubs on business trips and got in a round of golf where possible, not only as a company owner but as an employee. Despite the long work hours, we tried to make sure our lives were balanced.

A Vacation Home

The Siksika First Nation had built a vacation resort south of Cluny and Highway 1 on Alberta's Bow River. In September 1976 Nola and I purchased lot 27 on Crowfoot Circle. There were a total of 365 lots and a nine-hole golf course. It was a great "getaway" for us, and only seventy miles from Calgary. While in Banff attending the Doodlebug Golf Tournament, I noted a for-sale advertisement in the Calgary *Herald* for two Panabode log buildings. This was a real find for us as Nola and I had, for years, coveted owning a Panabode log house. Our first experience with Panabode log construction was in Dawson Creek where we often stayed at a motel constructed of such logs. The logs are dovetailed, one to the other, and are extremely weatherproof and warm. The interior cedar smell is very pleasant and never dissipates (the Japanese cornered the market on Panabode buildings from 1975 to 1995).

The homes were located at the trailer court in Bowden, Alberta, and one condition of the sale was immediate removal. On Monday, I contacted the owner and met him in Bowden. The larger building consisted of a living room, dining nook, kitchen, two bedrooms, and bathroom and was the office and living quarters for the motel owners. The smaller building consisted of three rooms: a separate shower and bathroom for women and men and a laundry room. There was an industrial-size furnace and heater in the basement and the house had a brick fireplace. I sold the furnace and boiler to a friend/customer. One of our part-time employees removed the fireplace, saving the bricks. I arranged for a moving company from Red Deer to transport the buildings to Siksika.

Our son, Brent, and Carla were being married on October 22 and Nola's and my twenty-fifth anniversary, complete with party, was scheduled for November 3. There was a lot of planning and work to accomplish in a very short period of time.

I contracted Nola's uncle Ed to supervise the foundation structure for the buildings. He was ideal for the job as he had been with the Federal Department of Indian Affairs on the reserve and knew who to hire to help with the bull work. I packed my tent and sleeping bag and spent a couple of days surveying the lot and determining where the buildings would be located. We were close to the river and there was a potential for flooding. With this in mind I laid out a plan whereby 8" x 8" x 6' cedar posts were sunk and cemented into the ground. This allowed free space from ground level to the floor of the house of nearly five feet, as 8" x 8" sleepers were fastened horizontally to the upright posts.

In the meantime Nola's mom and Hugh were staying at our home in anticipation of the wedding and anniversary. On the

Saturday of the building move, Hugh and I drove to Bowden to accompany the move. We arrived about fifteen minutes later than the time of departure and did not catch the trucks until Standard, some ninety-four miles down the road. They could not take the large building across the bridge over the Bow River but instead had to take it around through Lake McGregor and back north. Even with the detour, the buildings were on the foundation and the trucks left before 7 P.M. There was not a broken window or any damage to the buildings proper.

We restored the fireplace using river stone and used the fireplace bricks as walkways. We inserted a fresh air fireplace as we had in our Rockyford home that was 90 percent efficient rather than the norm, which in some cases is as low as 10 percent efficient. The fireplace was the sole source of heat for our vacation home. I ran into these fireplaces in Saskatoon. I was so pleased with the efficiency of the unit that I secured a distribution contract from the manufacturer. I immediately purchased six fireplaces, which I sold to other cottage owners at Siksika. We eventually set up Nola with a company to sell the units and we sold an additional number of fireplaces. Eventually, the work involved in selling and installing the fireplaces began to interfere with my commitment to Ace and we wound down the company. Nola bought me a beautiful bronze with the profits.

During this period, the Bank of Montreal was running a contest to determine the family that made the most deposits. As we were using the bank for Ace, as well as our personal banking, Nola suggested we enter. With the amount of money we were depositing, the bank gave us a pad of entries. We were advised that we had won one thousand dollars, which we used to build the fireplace. We removed a bank logo from an aluminum calendar and implanted it in the fireplace concrete.

We situated the buildings back to back and built a deck the full length on one side of the two buildings. In the "out building" we removed the plumbing in the men's shower room and installed bunk beds, providing a cozy guest room. The women's washroom was left for our second bathroom and shower. The other part of the "out building" was the washroom, where we installed a washer-dryer and workbench. We had enough room for storage of garden and other tools. Nola sanded and double stained the outer logs. We furnished the main building "cottage style," warm and friendly.

The grounds were contoured, with the house located on the upper level and a further two benches to the river. The first bench supported our garden and the lower bench sloped down to the river with a nice sandy beach. At night you could hear the beavers slapping their tails, warning their brood of impending danger. The grounds near the house were outfitted with a fire pit and horseshoe pits. There were some very ancient sweat lodge stones near one corner of the house. One day Nola was gardening near the stones and our dog, Deuce, started to growl and the hair on his back ridged. It was during the week and there was not a soul around our place. "Deuce, cool it," Nola told him. "It's only the souls of some Natives who are visiting the area." All of a sudden Deuce's rear end, including both legs, rose into the air and he let out a "yelp" as though he had been booted in the rear end. We invited some of the Siksika elders to come and have a look. They confirmed that this had been a very sacred sweat lodge. We believe! Do you?

The resort grounds included a lake and sandy beach as well as a super but difficult golf course, tennis courts, and of course the Bow River, where fishing was abundant. We had many parties, dances, and activities for the children. Our grandchildren enjoyed the beach and adjacent store, which sold ice cream and candy. There was also a restaurant and bar. Nola spent many hours during the summer with our grandchildren at the beach. She also started Jordan, Brent and Carla's son, on his golf career. She had read Harvey Penick's book on how to start young people. Using his suggestions, she would encourage Jordan to join her in putting ten balls from each station, gradually increasing the distance to the hole. The person who sunk the most balls per station received a nickel. Jordan turned out to be a scratch golfer and attended the World Juniors in San Diego. Nola and I played tennis with our grandson Jeffrey and she let the girls play "hairdresser," where our granddaughter Jennifer requested fifteen thousand dollars for one "hairdo." Sometime during the summer, we tried to have each grandchild out to the cottage for a one-on-one so that we could concentrate on his or her wishes.

Jim Munroe of the Siksika Nation told me that the resort area, which was in a horseshoe shape bounded on three sides by the Bow River, had been the location for the band cattle. He warned of the possibility of flooding and many years later we witnessed some flooding, but the devastating floods came after we had sold our interests at Siksika.

After we had been there for about ten years they decided to rename the resort Hidden Valley. I was the first Cottage Owner's Association president and thoroughly enjoyed dealing with the Siksika Nation executive. The properties were leased from the Band for forty years ending in 2013. Nola's and my attitude was simple: we would be well into our eighties at that time, so would not have any further use for the vacation spot.

The resort attracted a lot of press and our cottage was featured in a Calgary *Herald* article, particularly noting the Blackfoot name for our property, *Ni-Toy-Is*, which means "happy dwelling." Former premier Ralph Klein, while working as a TV reporter for a Calgary television station, visited the resort and that was the beginning of a pleasant acquaintance between us.

We paid a reasonable sum for the lease and were required to pay annual maintenance fees. As usual, in these matters, the maintenance fee was ridiculously low to attract purchasers. Of course, you had the purchasers who were not too savvy and never expected the fees to rise. That item alone caused a lot of grumbling but was unwarranted. The labour and maintenance crews were entirely Native and again some of our residents were not used to "Native

The Rintoul's first cabin, *Ni-Toy-Is*, at Siksika (Hidden) Valley
Summer Resort, was a Panabode log house.

time." For example, during Calgary Stampede, no golf or maintenance crew. That's the way it is; live with it. Eventually, the Nation
leased the entire property to the Home Owner's Association for a
fixed price, increasing yearly by the composite living index.

The water was turned on for April 15 and turned off October 15.
During this segment of the year, Nola and I lived at Hidden Valley.
I left the cottage at 5 A.M. for the office, driving with the sun at my
back. I was at the office by 6:30 A.M. and able to clear up my business with eastern Canada and finish my paper work by 9 A.M. I
completed the day by visiting customers and returned to Siksika
around 2:30 in the afternoon. I played a round of golf and relaxed.

After I retired in 1984, we began looking for a larger home at the
resort, as our plan was to live at Hidden Valley during the summer
and Arizona during the winter. We were fortunate to locate a two
storey, fully furnished house, Number 238, overlooking the second
fairway. Selling Number 27 was not a problem and this home was
large enough to satisfy our needs concerning sleeping accommodation for our ever-increasing family and all-round comfort. We added
a huge California cedar deck (around twelve hundred square feet).
Nola became known as "the woman of the big decks."

One night Nola was awakened by an unusual sound and asked

The second vacation home at the Siksika Resort was lot 238, 1990. The huge California cedar deck led to Nola becoming known as "the woman of the big decks."

me to investigate. When I entered the living room, the entire western sky was fire red. There was a fire in the cottage next to us and by now it was nearly destroyed. Fire brands were hitting our cedar shake roof and some fellows with a garden hose kept our roof watered down. As it was a particularly dry summer and we were surrounded by poplar trees, we decided to pack up and leave. We filled the Range Rover and our car in fifteen minutes with personal possessions and business computers and files. The fire did not reach us and we only suffered a burned hole in our roof, which was easy to repair. It took us the better part of a day to unload the vehicles and replace the goods, which demonstrates that under duress, the human body's adrenaline provides added strength and speed.

However, our plan to use Siksika as our retirement home in Canada was scuppered by the next-generation families who had replaced a lot of the old timers. Both parents worked in Calgary and often left their teenage children to fend for themselves during the week. This allowed the young folk access to the liquor supply and the raucous parties were too much for us. We sold, making a small profit, and concentrated our efforts on finding a Calgary retirement home located in a quiet area.

Sun City Winters

Our retirement home in Sun City, Arizona, was not only good for the arthritis, but grew to become a huge part of our lives. Nola and I wintered there for twenty-one years beginning in 1985 when we bought a sixteen-hundred square foot, one-bedroom townhouse with a two-car garage. My mother and father used our home in January and February each year and when Dad passed away in January 1990 we built an additional bedroom and bathroom for Mom. While she was able, she spent parts of each winter with us in Sun City. During this phase, we departed for Sun City after Canadian Thanksgiving and flew home for Christmas. Mother accompanied us on our return flight in early January and returned to Calgary after a month or so. We drove home sometime in the latter part of April.

Our home in Sun City provided us with the opportunity to bring the entire family there for Christmas. Individual family visits were also numerous and our grandchildren visited on an individual basis.

Skoki

A recreational and family highlight grew out of annual trips we began to make to Skoki Lodge in the Canadian Rockies. Each spring, for much of the 1970s, I led a group of family, friends, and customers on a cross-country ski excursion into the lodge, some thirteen miles north of Lake Louise.

It started when an employee suggested this beautiful spot for a cross-country expedition. He had skied into Skoki years earlier and used downhill skis with skins to climb the hills. I had taken up cross-country skiing to maintain my fitness during the winter months and this sounded like a good place to put my skills to the test.

The chap suggested that the trip into Skoki was relatively easy.

A winter home in Sun City, Arizona, became a place in the sun for the entire family.

Nola had also taken up cross-country skiing but was a raw novice. The trip involves a reasonably steep climb of about four miles to the upper Lake Louise Temple Lodge and then a continuous climb to Boulder Pass at fifty-six hundred feet above sea level, across Ptarmigan Lake and up Deception Pass at eighty-seven hundred feet, and a long gradual decline to Skoki Lodge at about seventy-two hundred feet. The ideal time to make the trip is late March or early April.

We parked our car at the Lake Louise Ski Lodge parking lot and started out about 9 A.M. Since we had been told it was an "easy ski" we took our time and opened our lunch bags at Boulder Pass where a relief cabin is located in case of problems. We sat in the snow and fed the whisky-jacks. It was a beautiful sunny day with a temperature of 5 degrees Fahrenheit. At those altitudes the sky was a deep azure blue.

After lunch we crossed Ptarmigan Lake and started up Deception.

The sun had been beating down on the east-facing trail and the skiing was extremely difficult. We changed our wax a few times and finally found one that gave us reasonable traction. On a difficult climb, I am the type who, once momentum is established, must keep going. Nola was having problems so I suggested that I would keep going and see her at the top of Deception.

On weekends there are many skiers who make the twenty-six-mile round trip in a day. They start out early in the morning, eat lunch at Skoki, and return in the afternoon. I asked one of the returnees to tell Nola, as he passed her, that I would wait for her. I can tell you that this comment did not appease her, as it took her a good two hours to arrive at the summit. She was very unhappy that I had left her on her own for the remainder of the climb.

We were now on the lee side of the mountain and it was late afternoon, with deep shadows and flat light. As I was relatively inexperienced in the essential nuances of cross-country skiing, I had not equipped my skis with safety harnesses. One ski came loose and sped off into a deep gorge about a half mile away. As the snow was twelve feet deep and I was on one ski it was impossible for me to rescue the wayward ski. So my dear wife traversed the steep incline on a back-and-forth route and returned with the ski. We arrived at Skoki Lodge after dark. It had been anything but "easy."

The lodge itself was delightful, built in the 1930s of logs complete with a large fireplace. There was a lean-to kitchen and a number of bedrooms on the second level and, of course, an outside biffy. They had added a few cabins since the first construction. Usually, a family leases the property on a yearly basis and provides linens, housekeeping services, and meals. The husband hauls supplies in and garbage out with a snowmobile and the wife provides the cooking and housework. The lodge is located very close to the headwaters of the Red Deer River and there are many adventurous trails located in the area that allow most pleasant day trips.

Nola insisted that I give her some lessons on cross-country skiing so she could comfortably return to civilization (an exercise like the blind leading the blind). The food and atmosphere at the lodge were fantastic and they have returnees each year from all over the world. I saw my first water oozle in a nearby stream. A water oozle is a bird that walks along the bottom of a stream hunting for tiny aquatic morsels of food.

That first trip into Skoki encouraged me to organize a yearly sojourn. Near the end of March I would take a group of about twelve to sixteen people, which filled the lodge. It was a very satisfying trip in and out and the stay for the weekend was an opportunity to get away from the toils of the city. They had all kinds of board games and the food was sumptuous and delicious.

Most of the gang returned to Calgary on Sunday, but I invariably stayed over until Monday. One Sunday evening, one of my customers suggested that Ace develop a surface explosive for portable seismic work. We examined the opportunity and later developed a very successful product that scooped our competition and provided us with handsome profits.

On one occasion when my son-in-law, John, and I were returning from Skoki on a Monday morning, we encountered a vicious blizzard on our way down Deception. Visibility was zero

Ski trips into Skoki in the Canadian Rockies in the 1970s with family,
friends, customers, and staff became an annual event.

and it was impossible to orient yourself. You could not tell whether you were going up, down, or sideways. We stopped for about half an hour to wait out the storm. John wondered what we would do if the storm did not subside. I suggested we would need to make a snow cave, crawl in, and wait out the storm. You had to be very careful, as the east side of Deception trail was avalanche prone.

On another occasion, one of the group and I were the first down Deception within a half mile of the lodge. We stopped for a rest and to wait for the others when one of my skis slipped and I fell, striking my shoulder on a large boulder (there was only about two feet of snow that year). A forty-pound pack on my back as well as my own weight resulted in a broken collarbone. I was in extreme pain and sleep was impossible. As I could use only one ski pole it was necessary for me to leave early on Sunday to avoid being a drag on the rest of the party. One of the fellows accompanied me to Lake Louise.

One other time on our way out my skis felt unusually heavy. When I looked back I saw that our dog, a Hungarian Vizsla who weighed seventy pounds, was standing on my skis and moving his legs in rhythm with mine. We all had a great laugh.

Terratima

Brenda, John, and some of their university friends organized a cross-country holiday over New Year's, 1976, at Terratima, near Rocky Mountain House. Larry and Clare Kennedy owned the ranch and also ran a few head of cattle there. They had moved in an old farmhouse and built a few cabins on the property as well as a sauna. There were a number of dozed seismic trails in the area and an abandoned schoolhouse nearby. Nola and I were the only parents invited, which we considered an honour. We attended for three or four years.

The days were dedicated to skiing as many miles as the human body could endure, and the evenings were spent playing games.

Charades became so intense that I commandeered my Price Waterhouse accountant to seal the movies or book names in an envelope. Nola and I supplied the New Year's dinner complete with turkey and all the trimmings. The other couples were responsible for other evening meals. The holiday usually spanned four or five days. The evening meal dish-washing chore was the responsibility of the men and was determined by a showdown with cards. The last two standing were responsible for the chore.

On New Year's Eve day, a group of us would ski out to the abandoned school, sweep the floors, remove dead flies from the window sills, cut some wood, and fire up the space stove. Larry would truck over cider to be mulled and the ingredients for bannock. There was an old organ and we danced and sang in the New Year.

An unfortunate outcome of our Terratima good times was the attitude of my father. The celebration of New Year's (Hogmanay in Scottish) is more important to Scots than Christmas. Since I was a little boy, I had accompanied my father on New Year's Eve visits to his Scottish friends' homes. It is a Scottish tradition that if a dark haired man crosses your threshold after midnight on New Year's Eve, and delivers a lump of coal, that household will have a prosperous year. My father was extremely upset because I had accepted my daughter and her friends' invitation rather than spending New Year's with him. When Dad died and we were going through his belongings we found a note attached to a gavel that he knew I treasured. In the note he referred to me as being selfish. This really hurt after all we had done for my mother and father. A couple of years passed before I realized that his actions were precipitated by me accepting my daughter's invitation rather than spending New Year's with him.

This situation gave me cause to examine my family's internal structure. I determined the most important beings to a mother and father are their offspring. Then those offspring have children of their own and their offspring become the most important beings to that generation. As a result, the original parents slip down one rung of the ladder in importance. When an original mother and father have great-grandchildren they cannot expect the same degree of attention that they enjoyed from their children, or their grandchildren. From my point of view, it is difficult for older people to understand this natural development.

For years I was angry with my father for the type of discipline he meted out and for accusing me of being selfish. My mother always said, "Oh Robert, Daddy was not very well. He did not mean what he said." I rejected this excuse. Nola believed that my father was jealous of my accomplishments. "After all, he did not reach

Evelyn and Robert Rintoul,
Lake Louise, Alberta, 1950.

his goals." This may be true and after a number of discussions with God, I finally forgave my dad. Ridding myself of those negative feelings has lifted a big weight off my shoulders.

Skoki and Terratima were not our only cross-country ski destinations. We skied Red Earth, Storm Mountain, and many other mountain trails. We spent many a pleasant Sunday after church at Christ Church Millarville skiing the foothills with Waverly (our minister) and Dorothy Gant and on some occasions with other parishioners. Fantastic Sunday evening meals at Dorothy and Waverly's followed the vigorous ski. We also accompanied Earle and Donna Mahaffy on their organized trips into Sunshine. They and most of their friends downhill skied and a few of us skied the cross-country trails. I also spent numerous pleasure-filled hours and days hiking trails in western Canada, accompanied by our son, Waverly Gant, and others.

Waterton

Brent and I hiked many pleasant trails in the Waterton Lakes area; two in particular stand out. These are Crypt and Twin Lakes. The Crypt hike was over a weekend and Jim Thompson, a friend and client of mine, joined us. I had a new pair of boots but had carefully (I thought) broken them in. It is necessary to traverse Waterton Lake via commercial boat to the west shore of the lake. It is a very steep climb to the falls, where we observed a mother water oozle nesting behind the spray.

My heels started to bother me. However, we continued on to the lake. We set up camp before I removed my boots and the sight was not reassuring. The heels on both feet were raw and bleeding. The weather turned ugly and it began to snow. By next morning our tents were knee-deep in snow. During the early evening, I observed a threesome—two fellows and a girl—tenting near us. Both fellows had cold weather gear but she was not so equipped and in fact was dressed in shorts and a T-shirt. I was appalled that neither fellow offered her any of their clothing. She was shivering profusely. Fortunately, I had an extra set of rain gear and offered it to her, which she gratefully accepted.

On Sunday morning my heels were no better and there was only one scheduled boat arrival. If we missed this sailing, we were obliged to spend another day on the west side of the lake. Jim offered to start out with me shortly after breakfast so I could slowly make my way down the mountain. Other than the heel problem, the trip was most enjoyable and the beautiful scenery made up for the inconvenience. Brent was to recover my "loaned" rain suit from the girl at a bar in Waterton, but that never occurred. The other trip to Twin Lakes was uneventful except for the fishing, which was very rewarding.

Another time, Nola and I were visiting Waterton and decided to take a day hike into Cameron Lake. We arrived around 1 P.M. and, as we had not packed a lunch, were ready for some food. Some previous fisher person had left a hook and a length of line snagged on a bush. I found an old piece of bread, which I used for bait, and hooked a good-sized Eastern Brook Trout. Nola found a piece of foil, I lit a fire, and we had our lunch. That was probably the tastiest fish we have ever eaten.

The Gold Trail and Other Adventures

One of the most interesting hikes I have ever been on was the Chilkoot Trail along which the Klondikers of 1898 travelled to the gold fields of Dawson City. Bob Leslie, who was the leader of a bar band in Whitehorse, Yukon, and myself decided to backpack the Chilkoot in July 1976. We boarded the White Pass & Yukon narrow gauge railway in Whitehorse and travelled to Skagway (a seaport where the Klondikers started their journey to Dawson City) and immediately started up the trail. Our route was exactly the one taken by the gold-hungry throng. We hiked from Skagway to Lake Bennett with the following way stations: Sawmill

Camp, Canyon City, Camp Pleasant, Sheep Shelter Camp, Summit, Stone Crib Emergency Camp, 21.5 Mile Camp, 23 Mile Camp, 25.5 Mile Camp (formerly Lindeman, abandoned), and finally Lake Bennett, thirty-two miles later. We covered the distance in three days and then boarded the train for Whitehorse. From here the Klondikers built boats, barges, rafts—whatever would float—and crossed Lake Bennett to Carcross and Tagish via lake systems to Whitehorse and the Yukon River to Dawson City.

Bob and Bob Leslie *(right)* at the Chilkoot Pass trailhead, 1976.

IN THE WORLD - MY ADRENILAN WAS FLOWING LIKE BLOOD - WITH MY VERTIGO PROBLEM I DIDN'T LOOK DOWN AT ALL.

- ONE FALSE STEP WITH A 40# PACK - NO MORE BOB.

- I BELIEVE WHEN I TOOK THAT PILE OF ROCKS IT WAS ONE OF THE HAPPIEST DAYS OF MY LIFE

- I WOULD NOT, I REPEAT WOULD NOT CLIMB THE CHILKOOT AGAIN - NEVER!

- AND TO THINK EVERY MAN, WOMAN & CHILD WERE REQUIRED BY THE RNWMP TO HAVE 1 YEARS PROVISIONS (ABOUT 1150#) AND ALL ONE COULD PACK WAS 150#

- THE RNWMP HAD A CUSTOMS OFFICE & A MAXIM GUN SET UP

NO FOOD NO GO - ANY BODY WHO TRIED TO SNEAK THROUGH WAS SHOT!

- HOW MANY MEN TOOK UP 3 OR 4 LOADS "SAID NO MORE" THESE ARE THE MEN WHO BUILT THE TOP - WAY & THE RAILROAD

- CAN YOU IMAGINE THE LOGISTICS IN 1897 - 98

- OK WE'RE AT THE TOP OF THE ROCKS BUT NOT QUITE AT THE SUMMIT

- ANOTHER STEEP CLIMB UP A GLACIER PAST A USA MONUMENT UP SOME MORE ROCKS & VIOLA THE SUMMIT SCENERY UNBELIEVABLE

- IT IS NOW 2:15PM WE LEFT AT 8:15AM FROM SHEEP CAMP

- 6 HOURS TO GO 3 MILES

- WE RESTED & GAVE OURSELVES A FULL

SAY J.C. WAS LOOKING AFTER US.

- I HOPE THE TRAIL NEVER BECOMES COMMERCIALIZED

- IT IS THE CONSENSUS OF OPINION THAT THE SUMMER TRAIL IS MUCH TOUGHER THAN THE WINTER TRAIL. AFTER ALL MOST OF THE KLONDIKERS MOVED OVER THE STREAMS & LAKES BY SLED ETC & WAITED AT LINDEMAN FOR BREAK UP.

- ALSO THE CHILKOOT PASS WAS COMPLETELY COVERED W/ SNOW IN WINTER SO NO CLIMBING ROCKS BUT W/ SLEDS & HAND RIDES ETC.

- I HAVE DONE IT, I WILL NOT DO IT AGAIN

- I TOLD BOB LESLIE THAT NOLA WOULDN'T TAKE $1,000,000 AND BRENDA $5,000,000 TAX FREE TO GO DOWN THIS TRAIL.

- THIS HAS TO BE THE MOST EXHILIRATING, EXCITING & INTERESTING ACCOMPLISHMENT CRR HAS EVER ENGAGED

WRIT BY HAND

Bob Rintoul

P.S. - THE U.S.A SIDE WAS MADE A PARK 2 WEEKS AGO. AND CANADA IN THE NEAR FUTURE

Another major hike was the West Coast Trail. Waverly Gant and I left Calgary on September 11, 1978. We drove to Victoria where a friend of Waverly's drove us to Port Renfrew and dropped us off. He then took Waverly's truck to Port Alberni and picked up his vehicle.

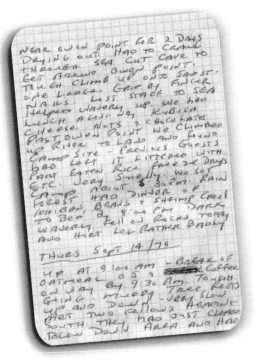

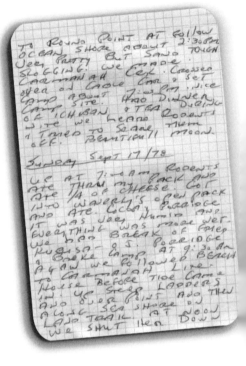

FLY TO WARD OF RAIN
NEITHER WERE VERY EFFECTIVE
WE WENT TO BED WITH NO
DINNER. BY MORNING OUR
SLEEPING BAGS, OURSELVES
PACKS CLOTHES, EVERYTHING
WAS SOAKING WET.

Friday Sept 15/78

UP AT 7:00 AM SOAKING
WET BREAK CAMP ON
BREAK OF CHEESE &
KUBASA. WAVERLY HAD
A GREEN ONION OR TWO
WHEW! PACKS NOW
WEIGH ABOUT 65#. IT
HAS QUIT RAINING BUT
4 WEED GROWTH IS SOAKING
& PIT JEANS IN PACK AND
ONLY WEAR RAIN SUIT.
THEN SCRATCH CRAWL
ETC DOWN TO SANDSTONE
OAK AND SCRATCH CRAWL
YE FORTUNATLY WATER
CLOSS OVER SAND STONE &
DOWN FALLS MY SOLE
DEEP. THEN DOWN SCRATCH
CRAWL ETC INTO CULLITE
CAR. AND LOW AND
BEHOLD THERE IS SAME
PARKS CREW WOMEN-
CING ANOTHER CABLE

TO ROUND POINT AT FOLLOW
OCEAN SHORE ABOUT 2:30PM
JEER PRETTY BUT SAND TOUGH
SLOGGING. WE MADE
CAARMANAH CEK. CROSSED
OVER ON CANOE CAR & SET
CAMP ABOUT 7:00PM. NICE
CAMP SITE. HAD DINNER
OF ICHIBAN & TEA. DURING
NITE WE HEARD RODENTS
& TRIED TO SCARE THEM
OFF. BEAUTIFUL MOON.

Sunday Sept 17/78

UP AT 7:00 AM RODENTS
ATE THRU MY PACK AND
ATE 1/4 OF CHEESE. GOT
INTO WAVERLY'S OPEN PACK
AND ATE LOCKY PORRIDGE
IT WAS WET. HUMID AND
EVERYTHING WAS MORE WET.
WE HAD BREAK OF FRIED
KUBASA - O.J. PORRIDGE
& BRAKE CAMP AT 9:30 AM
AGAIN WE FOLLOWED BEACH
TO CARMANAH LINE
HOUSE BEFORE TIDE CAME
IN. UP STEEP LADDERS
AND OVER POINT AND THEN
ALONG SEA SHORE OJ
LAND TRAIL AT NOON
WE SHUT HER DOWN

Bob in Scottish gear. Bob and son, Brent
(pictured here), visited Scotland in 1992
to explore their heritage.

Anyone who knows me understands that golf and Scotland are both passions of mine. In 1992, I was able to combine the two when I took Brent to Scotland with me to play in the Father and Son Golf Tournament. We played two rounds at Turnburry, one round each at Old and New St. Andrews, Old Prestwick (site of the first British Open), and Carnoustie. We came second, and won hand-made wooden putters from Auchterlonies shop in St. Andrews. After our golf we toured Scotland for another couple of weeks. It was an opportunity for Brent and me to discover our Scottish roots.

Hunting and Farming

Over the years, Brent and I spent many delightful hours during the fall, hunting upland birds. Many times we included customers, and on a few occasions rented a trailer unit for sleeping, consuming a few libations and cooking a meal or two. Our dog, Deuce, was a real asset. He pointed, flushed, and retrieved all manner of upland game birds with ease and had a pleasant personality. He was also a great family dog.

There are many stories about Deuce. Our next-door neighbour in Haysboro located their patio between our homes, and when they had company, Deuce would crawl up the fence, hook both forelegs and feet over the top of the fence and join the party. On one hunting trip he was working for our party of six hunters. As we headed back to the trailer late in the evening he was loping ahead of us rather than flushing the ditch. I gently pushed him into the ditch and he immediately came back up on the road and looked back at me as if to say, "I'm tired. That's it for the day."

I established a trap-shooting area in the northwest corner of our land. At one time there had been a slough surrounded on two sides by shimmering aspen and low bush with nice thick grass in the middle. It was about a half acre in size. I bought a used twenty-five-foot trailer complete with awning, etc. Immediately before upland bird hunting each year, I would invite customers to

Trap shooting with customers at Rockyford farm, 1980s. Hunting with his son, Brent, dog Deuce, and customers was a way for everyone to unwind.

sharpen their accuracy with a shotgun. I supplied the shells and a double trap thrower. We would shoot in the afternoon then have a few drinks and I would cook up a steak dinner. After dark we would proceed to the house and shoot pool until midnight. It was a very relaxing afternoon and evening for our customers after the hectic pace of the city.

On one of our pheasant hunting trips our luck was miserable, and we saw some ducks in a slough. Deuce was not a waterfowl dog, so we left him in the car. When we returned, he had the driver door upholstery in shreds and the wing window latch open. I was furious, and said, "That's it; the dog goes." Brent intervened and offered to find parts and fix the mess. When he completed the repair job, you would never have known it was damaged. In fairness to Deuce, when the four of us left the car with our guns, he must have felt we were shunning him.

Deuce loved attention. While hunting in harvested canola fields, the stocks are about inside-leg high on a dog and are very sharp and pointed. After a day in the harvested fields, his leg pits would be sore and raw. When we arrived home after such a day, Deuce would head for Nola's and my bedroom and start to whine. When a family member entered the bedroom, he would have one leg up on the bed looking at us in this helpless manner until we helped him raise the other leg. Normally, it was an easy hop onto the bed.

Bob with the family dog, Deuce. Ace Explosives facility near Rockyford, Alberta, is in the background, 1980.

The evening before opening day of pheasant season, Brent, the dog, and I would take our station wagon and park near a likely spot for an early morning hunt. We had a foam mattress in the back of the wagon and a couple of sleeping bags. The dog was supposed to sleep at our feet but would work at forcing himself into one of our bags. If we did not relent he would start to work on the zipper. He had learned to open doors at the house by knocking out the facing plate on a doorknob, inserting his paw and turning the knob, thereby entering the room.

Nola decided she would like to raise pheasants to replace the ones that Brent and I shot. We built a chicken-wired pheasant run down the hill from our Rockyford home, approximately one hundred feet by twenty feet and installed the wire to a depth of about two feet below ground level. We acquired fifty hen chicks from the Alberta Pheasant Farm in Brooks. We fed and watered them until they were about twelve weeks old, then released them into the wild. It was interesting to see the wild cock pheasants gather and round up four or five hens as their mates.

I talked to the owners of the land surrounding us and convinced them to let me establish a game preserve. I procured letters of permission from each one, and, due to our explosives storage, patrolled the perimeter during hunting season. Our land had not been fenced for years and was open to transgressors. I enjoyed fencing and cross-fencing the property.

One of the benefits of our country business location was the opportunity to test my skills as a farmer and rancher learned at my grandfather's knee. I contacted the local Alberta District Agriculturist and he told me that we had about one hundred acres of original prairie grass that was loaded with nutrients. He cautioned me to wait until June 1 to put our cattle out to pasture and remove them by October 1. This would allow our pasture grasses time to renew and we would continue to be blessed with original prairie grass. We followed his advice and after removing our herd from open grazing, finished them with grain until early November. We never used growth hormones and had a waiting list of Calgary purchasers for our beef.

Our first year, we purchased fifty head of yearlings from the short-grass areas near Bassano. They were a wild bunch, as they had not been near humans all winter. A shirt-tail relative of mine suggested that we invest in llamas. We purchased one male and two pregnant females. We kept the male and one female at our place and one female at his farm. We put the Bassano cattle through dehorning and ear tagging and let them loose into a holding pen. When they saw the llamas, they headed west and took out every fence for two miles. Fortunately, we had a couple of cowhands on payroll, but we still had a great rodeo. We finally caught the last animal about 7 P.M. that evening. The cattle got used to the llamas and we had a reasonably quiet summer.

About November 1 we slaughtered the cattle at the Rockyford abattoir, hung them for twenty-one days, and offered sides of beef (butchered to their specifications) to our Calgary customers. After supplying free beef to all our family, we still made a handsome profit. In the meantime our llama herd was growing.

We planted the northwest quarter in alfalfa and another quarter in winter rye (at the agriculturist's suggestion). We took off two cuttings of alfalfa and our winter rye yielded sixty-five bushels to the acre. We purchased and I operated the necessary farm equipment and we operated the grain on a partnership basis with a local farmer. We also had chickens and geese. It provided us with enjoyable summers and offered me an opportunity to prove my farm worth.

The following summer, a friend of mine offered me the opportunity, at an excellent price, to purchase ten head of purebred Maine Anjou cows from him and his accountant. Apparently, they had purchased these animals as a tax write-off, which had not materialized. Half were at a feedlot south of Claresholm and the other five were at a cow-milking farm west of Balzac. The ones at the milking farm were in very poor shape, with pink eye and warbles. It did not take us long to clear up their afflictions and start to produce a very healthy herd.

Maine Anjou are a cross between the British Maine, which produces large animals, and the French Anjou, which are excellent milkers. In other words, they are the best of both worlds. It was necessary to breed the yearlings back to a Red Angus, which is a smaller bull. Otherwise we would have encountered birthing problems. We kept proper breeding records and culled inferior yearlings. We butchered the culls and sold the beef to our Calgary customers as well as providing the family with meat. After the first year of breeding the yearlings and our original cows, we were able to employ a Maine Anjou bull. I purchased one of a set of twins from a seismic customer and friend. I knew from my investigations that it was probable that one of the twins might be sterile. After a missed cycle on the cows, I asked Doug if he had tested the animal and he said no. We shipped the dud bull to the meat plant and Doug guaranteed any out-of-pocket expenses. He immediately purchased and shipped another Maine Anjou bull that ably performed his duties. The new bull was a very friendly animal, and although

Fieldstone Farms was a working farm, and even included a small herd of llamas. Pictured here are Push Me, Pull Me, and Little Ace (the baby), 1984.

a huge beast, would come over to the corral fence and allow me to scratch his forehead. Thereafter we settled into a pattern of breeding our yearlings to a Red Angus and our cows to our Maine Anjou bull.

Our animals were very sleek and well formed because of the excellent feed and water. We named the animals and treated them like humans. I had placed a water trough just down hill from the house and one of my great delights was to arise about 5:30 A.M., grab a cup of coffee, and sit on the deck. I could hear the cattle munching down a coulee toward the water. They would arrive and our German Short Hair pup would stand on one side of the fence and have a discussion with a particular yearling, whom we named Freckles.

Our llama herd grew and after a couple of years my partner asked if I was interested in buying his half of the herd. I concurred and continued to raise and improve the herd. Eventually a farmer from northern Saskatchewan arrived at the farm and offered us forty thousand dollars in grain for the herd of about ten. I declined, and he came back the following spring to pay cash and the price increased to forty-five thousand dollars. He issued a certified cheque and loaded the herd in his cattle trailer and away he went. Our timing was impeccable (just lucky), as the Canadian Government discontinued the opportunity to import llamas from South America (primarily Peru). We tripled our investment in three years.

What does a person do with llamas? They may be shorn for their wool, butchered for meat, or used by sheep farmers to deter coyote invasions. However, at the time we were involved, they were pets for the rich and famous, primarily in California. Llamas make great pets and they loved our grandchildren, allowing them to pet them.

Llamas only spit when annoyed. A female llama determines her own ovulation period, and the gestation period is close to one year. As soon as the female drops her offspring, the male wants to breed her. If she does not agree with his advances, she lets him have it with a big gob of spit. He quickly gets the message! The birthing process is a social event and the entire herd gathers around the birthing female. As long as I approached a new birth site on hands and knees, they would allow me to examine the newborn to determine its sex. Llama inter-herd communication is issued by a series of "hums." I became quite adept at conversing with our llamas and on occasion when visiting a zoo, entertained the children. We have witnessed adult llamas teaching their young in the art of warding off coyotes.

We also pastured two or three horses. I purchased a quarter horse mare for myself and an Arabian for Nola. One of our employees was given an older horse and our farm became the animal's retirement home. My mare was out of Tinky Poo, an excellent quarter horse mare, but she was cantankerous. I finally gave her to one of our cowboys to train. She was fine for about a month but then reverted to her old self. I finally kept her pregnant and barefoot! Jake, the horseman at the Stahville Hutterite colony a couple of miles north of us, had a beautiful stud. I had her bred every year, sold the foals, and finally sold her at our closing auction sale.

The farm became a place for their grandchildren to learn about life in the country. Here Bob spends time with his granddaughter Heather, 1980.

Rintoul family Christmas at Fieldstone Farms, 1985.

The farm was a great learning experience for our grandchildren, where they were able to mingle with all kinds of domesticated and wild animals and birds including deer, coyotes, porcupines, skunks, badgers, wild mink, pheasants, great horned owls, red tailed hawks, and more, too many to mention. At my urging, Ducks Unlimited built a couple of dams on our property and during the winter I cleared the snow off the ice and provided lighting and music for skating parties.

Christmas time was a lot of fun, as the entire family would gather at the farm. On Christmas Eve, the children went to bed and I would hover outside with a set of sleigh bells, jingling away, and in a very deep voice, imitate Santa. I could hear them cautioning each other to get to sleep as Santa was arriving.

When our family visited during the summer we would set up and ignite fireworks, which were not allowed in the city. We had built a sandbox into the deck and provided toys and blocks for the younger grandchildren.

Those were good years.

Charity Begins at Home

I have been blessed with the opportunity to experience three major epiphanies in my life. The first one was in my early thirties when I realized that my mother had "made peace with herself" when I was just a child. That was probably why she got along so well with my dad. My definition of "making peace with yourself" is when one day you look into a mirror and say, "I love myself and therefore the only person I must please is me." There is a caveat to this explanation: you must not harm other people with your decisions or actions. At that point in your life you do not need to attempt to please everybody, particularly to your own detriment. The second epiphany was the day in my late thirties when I was able to look in the mirror and make the same declaration. Maybe I knew it all along but was unable to define it until I recognized my mother's situation. If you try to please everybody you are inevitably going to prostitute your principles. Making peace with oneself is essential to success in life.

My third epiphany was in my late fifties when I received an answer to a long-standing question about myself. I had great difficulty understanding why whatever intelligence and monetary success I had accomplished were greater than a large number of my school buddies, hockey compatriots, and business associates. These folks, in most cases, had a better education and as many opportunities to grab the "brass ring." One day, while meditating about a number of life's questions, I heard a voice say, "Robert, whatever success you have attained does not belong to you in its entirety. You are only a steward of these attributes and must share all you have with the less fortunate." At that point I realized that a person does not accomplish their success by their efforts alone. You benefit from some other force "out there." This force may be whatever name you wish to call it. In my case, it is God's assistance. I listen to and ask help from Him every day. This epiphany started my philanthropic drive, both financially and morally. Not only assisting with money, but also attempting to help young people and others who wish to mine my advice while struggling with their own endeavours.

There is a saying that "charity begins at home." Many years before my third epiphany, Nola and I decided to help our family. We did not gather any wealth until after we had bought out our backers and paid off the bank. We had taken over Brent and Brenda's mortgages and charged them a much lower rate of interest. On our way to Sherwood Park to join Brenda and her family for Christmas, I suggested we pay off the children's mortgages. Nola agreed, and as the outstanding mortgages were similar in value, it simplified the process. They were elated with their Christmas present.

We reasoned that we could live to a ripe old age and whatever wealth we would accumulate would pass to our family when they were too old to use the funds. With that in mind we instituted yearly funds for both families, suggesting they use it for upgrades, etc. We also set up education funds for our five grandchildren, adding annual amounts. We have continued education grants for our great-grandchildren. When we sold the company, we gave a trip and a sum of money to my parents (Nola's parents had passed away), as well as both families. When our grandchildren married, we offered to bankroll them on their first mortgages thereby allowing them to purchase a house rather than waste their money on rental accommodations. These funds were set up as a draw against their inheritances.

My mother and father were living in an apartment on Heritage Drive. The building was going downhill and a number of the other renters were of questionable character. There was no underground parking and during the winter my mother had to sweep the snow off the car and on occasion local kids would pull out the block heater cord. Both Nola and I were concerned and decided to augment their living costs. We eventually purchased a condo, to their satisfaction, on Twenty-sixth Avenue and Fourth Street SW. It was located on

Family photo, 2007. *(Back, L – R):* Jordan, Jennifer, Joe, Heather, Josh, Susan, Jeffrey. *(Front, L – R):* Brent, Carla, Nola, Bob, Brenda, John.

Bob and Nola's great-grandchildren, 2009. *(From left):* Sophie, Jacob, Heather, and Evelyn.

the eighth floor of the Riverstone building, overlooking the city, and two blocks from where I was raised as a teenager. Of course, there were maintenance fees, and we wanted them to feel comfortable and maintain their dignity and self-worth. They paid whatever contribution they felt comfortable with, and it worked well. After Dad died in 1990, Mom continued to live at Riverstone until 2000 when we moved her to Trinity Lodge, which afforded full assisted living until she died in 2003 at age ninety-four.

We also provided a special trip of their choice on their fiftieth birthday for our children and their spouses. John, Brenda's husband, was first, and I accompanied him to the Daytona 500. Nola accompanied Brenda on a special trip to Moose Jaw, taking advantage of the spa and other tourist attractions such as the tunnels and wall murals. This was also a nostalgic trip for Brenda, as her grandmother had lived in Moose Jaw. Brent and I took in the Masters Golf Tournament in Augusta, Georgia. Last but not least, Carla, Brent's wife, and Nola visited Las Vegas, and among other highlights attended Celine Dion's show at Caesar's Palace.

It made sense for us to start a program of philanthropic endeavours. It started in 1981 with the Alpha project that financed the cancer wing at the Foothills Hospital in Calgary. We pledged over five years and provided a lab in the name of Bob and Nola Rintoul. The following year Nola had a spinal fusion. It is difficult to express the joy we felt when I wheeled her down to have

Four generations golfing at Sun City, 1997. *(L – R):* Jordan, Brent, Bob, and Bob's mother, Evelyn, who took up golf at age forty-six.

our first look at the lab. Subsequently, Dr. Brown, an oncologist at the University of Calgary, used the lab as his office. He contacted us and thanked us profusely for our donation. That gesture added to our joy in providing the lab.

Our philanthropic endeavours continued in 1983 with the provision of computer upgrades over a period of five years to a top cancer-research scientist lab at the U of C. Taiki Tamaoki, originally from Japan, had been recruited to a large cancer research facility in California. Subsequently, the U of C recruited him for their program, with hopes of solving the origin of this terrible disease. Tamaoki had purchased a quarter section of land next to ours at Rockyford. Forty acres of this land were of no use to him as they were located in a coulee, where our cattle grazed. After a couple of years we negotiated the purchase of the forty acres. While discussing his work at the U of C, I became aware that he had to lecture each year to procure enough money to fund his computer upgrades of five thousand dollars per year. This did not seem right to me—a scientist of his stature wasting valuable research energy on raising funds! We set up a five-year program, and as Nola's mom and stepdad had died of cancer, we named the fund in their honour, Victoria and Hugh Brown.

Our next "give back to society" involved establishing a scholarship for a medical doctor on the Siksika Reserve near Gleichen, Alberta. It came to our attention that a non-Native doctor was visiting the reserve every couple of weeks to attend to the local patients. We quickly concluded that a doctor without an understanding of Native culture was not the answer to the reserve's requirements for healthcare. With this thought in mind, in 1992, we established a five-year medical scholarship for a reserve doctor who in some way was affiliated with the Siksika Nation. The Elders named the scholarship "Sookinaksin," which is Blackfoot for "healing." The Nation established the parameters for the scholarship, including Siksika membership and years of service required to the Nation for each year of scholarship. Finally, in 2007, a successful candidate applied for and received the scholarship now funded at thirty-five thousand dollars.

Or next commitment was to Fort Calgary. As a born and raised Calgarian, I am very interested in our history. We had bought an "inch of land" many years ago to help a previous Fort Calgary fundraiser to rebuild the fort. There was a hiatus of many years, until 1998, when they contacted us for a sizable donation to help rebuild the area with a proper interpretive building, etc. On reviewing the plans, I noted the inclusion in the rebuild of the original Calgary Electric Street Car Railway scenic car. This car was built in 1909 as a way to transport investors around Calgary to view property. It was open on top and had mirrors and scenes painted on the sides. The seating was arranged so that it sloped up slightly from front to back. During World War II it was resurrected and used for the sale of V (victory) bonds. The donation was in the Silver Sponsor category, and as my father had been a mechanic at the Calgary Electric Street Car Railway, it was dedicated in his honour. The replica car is situated at the beginning of the Street of Dreams, and a large television monitor is located where the conductor would normally stand. When a "passenger" enters and is seated, the monitor depicts the story with photos of Calgary in 1909. There is a plaque mounted near the entrance to the car noting the dedication in my father's name and some written and visual history of his days at the street railway. The Street of Dreams walks the viewers past the *Herald* typesetting room, the Palace Theatre, a drug store, and more.

Nola and I felt we would like to contribute something for osteoarthritis in bone and joint research. Nola had a hip replacement in 1990, and I was in bad shape with both knees due to osteoarthritis. Dr. Glen Edwards was our orthopedic surgeon, and one day in October 1999, while discussing my ailing knees with him, I asked him what we could do to further research in the area of osteoarthritis relating to bones and joints. Edwards suggested we contact Dr. Cy Frank, MD, FRCSC, McCaig Professor in Joint Injury and Arthritis Research; Professor, Department of Surgery; Chief, Division of Orthopaedics, University of Calgary and Calgary Regional Health Authority. Dr. Frank is the recognized expert in Bone and Joint, not only in Calgary and Alberta but federally. He was appointed by the Canadian Government to head up a five year, 87-million-dollar program to improve research of bone and joint problems. Dr. Frank suggested we set up an endowment fund to the McCaig Centre in Joint Injury and Arthritis Research.

We agreed and established the Gordon R. and Nola E. Rintoul Endowment Fund dedicated to supporting arthritis research through the McCaig Centre for Joint Injury and Arthritis Research. Currently (2009) the market value of the endowment is $331,000. The U of C wanted to hold a presentation announcement party for family, friends, and luminaries in the medical field.

Calgary's Scenic Car, 1924. The Rintouls' donation to Fort Calgary helped keep this scene alive in the memory of Calgarians. Bob's father worked for the Calgary Electric Street Car Railway for much of his life. He is remembered with a plaque at the scenic streetcar at Fort Calgary.

We declined, as we did not feel comfortable with this method of announcing the endowment or with any other publicity.

They asked if they could place a plaque in the boardroom at the McCaig Centre. We agreed, and they invited us to lunch and to unveil the plaque. To our surprise, they had commissioned a U of C arts student to paint a prairie scene on canvas, mounted with a brass plate, with the particulars of the endowment. They provided us with a duplicate for our home. We were overwhelmed.

The dean of medicine, Grant Gall (deceased April 2009), hosted a lunch at his favourite Italian restaurant for Nola and me and others from the university. Nola and I have exclaimed many times that our affiliation with the university has been a win-win situation. We were able to provide funds to help humanity, and we in turn met many interesting people at the university. As a person ages, your circle of friends and acquaintances, due to deaths, grows smaller

PHOTO
COLLINS. 1924.

and smaller. This situation introduced us to an entirely new group of fascinating people. As neither Nola nor I attended university, it opened up a world of interest we had not had the opportunity to enjoy. Grant Gall and I discovered we had a number of similar interests and became close friends along with his wife, Lori.

After selling a quarter section of land near High River and disposing of our winter home in Arizona, we decided to further our contributions to Bone and Joint Research. In the fall of 2006, we again approached Dr. Frank and divulged the amount of funds we wished to donate. The amount we were prepared to pledge took him by surprise and he immediately suggested an Endowed Research Chair in support of the Alberta Bone and Joint Health Institute. A chair at the university level requires 3 million dollars. Fortunately, the provincial government matches funds donated. Once all the funds are in place, the U of C will place applications all over the world to attempt to recruit a top research person for the chair.

After giving the situation a lot of thought over the past two years, it surprises me that this will be the first chair for Bone and Joint when you consider the amount of funds expended yearly on patients afflicted with bone and joint disease. The McCaig endowment set up in 1999 will eventually be transferred to the chair.

In this case the university was determined to publicly announce our gift. We acquiesced and the announcement function was held at the Ranchmen's Club on May 11, 2007. As Ken King was co-chair of Reach (the financial administrative arm of the U of C for donations), the university approached him to chair the announcement of the donation. He refused at first, on the grounds that he kept Friday evenings open for his family. But when he learned the identity of the donors he agreed to the task. On completion of the "speeches," Ken presented us with a silver hockey stick appropriately engraved with the details of the contribution (Ken is president of the Calgary Flames). We were thunderstruck. All NHL players achieving one thousand games are awarded a silver hockey stick. Ken had sought out special dispensation from the NHL to award

Calgary Flames president and co-chair of Reach, Ken King, presented a silver hockey stick to the Rintouls on May 11, 2007, at the announcement of a Research Chair for the Bone and Joint Institute at the University of Calgary. The hockey stick holds a place of honour in the Rintoul home.

a silver stick outside the league. It will likely be the only one so awarded. It is displayed in a place of honour in our trophy room.

I am, due to my great-grandparents homesteading in southern Alberta in 1889, a member of the Southern Alberta Pioneers and their Descendants (SAPTD). To qualify as a member your ancestors must have landed in southern Alberta by midnight December 31, 1890. I eventually talked my mother into joining, at which time I became more active in the organization. As you can well imagine the majority of members are seniors. The SAPTD building is located on land leased from Calgary, on the hill at the far south end of Fourth Street SW, overlooking the Elbow River.

The parking lot had a gravel top, and I became concerned that one of our members would fall and break a hip or be otherwise injured. I proposed to the board that Nola and I would like to build a proper paved parking area. They accepted and I commandeered an old school friend of mine with construction experience, Bob Boon (since deceased), to quarterback the construction. It took four years to get the approvals from the City departments, but it was finally completed in 2004. We dedicated it in my mother's name and a plaque is affixed to the cairn on the grounds.

We contribute one other small donation to Hull Child and Family Services. We provide two season tickets to the Calgary Flames games. They distribute the tickets as an incentive to the young residents. Each year we receive many handwritten or printed "thank yous" from the young people, expressing their thanks. The comments on these cards are heart warming.

A trio of minor contributions consisted of a brick in each of our five grandchildren's names to pave Olympic Plaza, a donation to the Cross Canada Trail, and three audience chairs at the Performing Arts Centre in Calgary.

Further contributions have been made to Lougheed House, where a bench was provided in my mother's name and a contribution in time and money to the "Egg Money" project. This was the brainstorm of Don and Shirley Begg. They own and operate a bronze foundry and studio, Studio West Ltd., in Cochrane, Alberta. Shortly before she passed away, Shirley's mother left a sum of money to recognize Alberta's pioneer women. She requested: "We're always recognizing the deeds of the pioneer men and not the ladies. Please correct this oversight."

Shirley and Don (his sculptures are well known throughout the world; some adorn the Calgary International Airport) decided to develop a statue of a pioneer lady with her two young children gathering eggs and feeding the chickens. They are both shy people and required help to get this project off the ground. They hooked up with the Ranche at Fish Creek Restoration Society, which, along with others, pushed the project to completion. They raised the necessary funds by asking for five-thousand-dollar donations. This entitled the donor to a maquette of the statue (table-size bronze version of the original) and the opportunity to dedicate their contribution to a pioneer lady. This dedication would consist of a plaque at the base of the full-size statue.

I recognized my grandmother, Edith Louise Burke, my mother's mother. I suggested Mother's Day, May 12, 2002, as the date for the dedication, and that was accepted. Her Honour, the

The "Egg Money" statue by Cochrane, Alberta, sculptor Don Begg was commissioned to honour the contribution of pioneer women. Bob and Nola dedicated their donation to Bob's grandmother, who, along with her young grandson Robert Rintoul, candled and sold eggs to Calgary homemakers.

Lieutenant Governor of Alberta, the Honourable Lois E. Hole, CM AOE, unveiled the statue. Who better than Her Honour to unveil the statue—an Alberta pioneer woman in her own right? The statue is located a little west and north of the Ranche restaurant in a park planted with native greenery. The reason I honoured my grandmother was because of her dedication to her family and her egg candling at the farm.

I have been fortunate enough to have the opportunity to serve on many company and organizational boards of directors. I started this type of volunteer work in 1971 when I was elected secretary of the Canadian Society of Exploration Geophysicists, followed by Heritage Park, the Alberta Motor Transport Association, Bank of Montreal Small Business Advisor for Western Canada (charter member), Canadian Explosives Distributors Association (founder and first president), Nobel Insurance, the Alberta Motor Truck Association, the Group of Twelve (who bankrolled the Olympic Hockey Club move to Calgary in 1978), XL Beef, Southern Alberta Pioneers and Descendants, Petroleum History Society (vice-president), and a member of the Calgary Flames Hockey Club Ambassadors.

The most satisfying board experience came through my involvement with the Canadian Federation of Independent Business (CFIB). Our president and cofounder was John Bulloch, who was well placed to take our cause to the federal government. At that time—1979—we represented more than sixty-six thousand small business organizations with fewer than twenty-five employees. Most Canadians do not realize that these small business companies command more than 90 percent of the total Canadian work force. It gave me a great deal of satisfaction to have the opportunity to contribute to Canada's welfare and to assist in the formulation of government policy from the "back room." Mr. Bulloch and I had some grand arguments along the way, as he was prone to stating, "My way or no way!" ∎

Bob and Nola as pictured on the announcement of their donation to
the Alberta Bone and Joint Institute, 2007. Still dancing.

Conclusion

As the obituary I have prepared says: I have no regrets and no apologies. Any apologies I have been required to make, I have made to the aggrieved person at the time. I have never sacrificed my principles to satisfy other people.

Nola has never second-guessed me on any business deal. It was not often that I asked her about a decision, but if the deal affected the family, then the whole family was involved—Nola as well as the children, and we came to a consensus. If Nola did not agree with what I wanted to do she would tell me so. She would present her arguments, and invariably I acquiesced to her position because if she felt strongly enough to talk to me about it, I knew it must be important to her.

I have often thought that I would like to learn to fly an airplane, but I knew better. I'm a risk-taker. I would fly along the wrong side of a cloud and say: "I can make it through that cloud." And I would not be here today. I recognized that tendency and chose to do other things instead.

I've lived a full life. I try to make life a little better for others as well as our family, and that pretty much summarizes my time on this earth. I've done my best.

Index

About the Authors

Gordon Robert (Bob) Rintoul was born in Calgary, Alberta, in 1930 and completed his education at Calgary's Central Collegiate High School. He married Nola E. Cooper of Moose Jaw, Saskatchewan, in 1951, and they now have two children, five grandchildren, and four great-grandchildren.

Bob worked for Safeway as a young adult, and joined the Geophysical industry in 1950, attaining the position of party manager. In 1957 he began work in the explosives industry and in 1967 started his own explosives distribution business, Ace Explosives Limited, distributing for both Canadian Industries Limited and subsequently DuPont of Canada Limited in western Canada and the Territories. His employees bought him out in 1984 at which time he retired.

While in business Bob served the community in the following organizations: Secretary of the Canadian Society of Exploration Geophysicists; Alberta Motor Truck Association Board; Calgary Heritage Park Board; Charter Member of the Bank of Montreal Small Business Advisory Panel; Co-founder, first President, and Honorary Life Member of the Canadian Explosives Distributors Association of Canada; Member of the Board of the Canadian Federation of Independent Business; and a member of the Group of Twelve, who made it possible for the Canadian National Hockey Team to move to Calgary in 1978.

Currently, Bob is on the board and is Vice-president of the Petroleum History Society, a member of the Southern Alberta Pioneers and Descendants, an Ambassador for the Calgary Flames, and on the advisory boards of the Alberta Bone and Joint and Patient Wellness for the new Calgary Health Region South Campus Hospital.

David Finch researches and writes the history of the Canadian West and is the author of more than twenty books, including *Pumped: Everyone's Guide to the Oil Patch.*